Jean-Pierre Serre

Lectures on the Mordell-Weil Theorem

Aspects of Mathematics

Aspekte der Mathematik

Editor: Klas Diederich

All volumes of the series are listed on pages 219–220

Jean-Pierre Serre

Lectures on the Mordell-Weil Theorem

Translated and edited by Martin Brown
from notes by Michel Waldschmidt

V

Springer Fachmedien Wiesbaden GmbH

CIP-Titelaufnahme der Deutschen Bibliothek

Serre, Jean-Pierre:
Lectures on the Mordell-Weil theorem /
Jean-Pierre Serre. Transl. and ed. by
Martin Brown. From notes by Michel
Waldschmidt. – Wiesbaden; Braunschweig:
Vieweg, 1989
(Aspects of mathematics: E; Vol. 15)

NE: Waldschmidt, Michel [Bearb.];
Aspects of mathematics / E

Prof. *Jean-Pierre Serre*
Collège de France
Chaire d'Algèbre et Géométrie
75005 Paris

AMS Subject Classification: 14 G 13, 14 K 10, 14 K 15

ISBN 978-3-528-08968-9 ISBN 978-3-663-14060-3 (eBook)
DOI 10.1007/978-3-663-14060-3

Originally published by Friedr. Vieweg & Sohn Verlagsgesellschaft mbH,
Braunschweig in 1989.

Produced by Wilhelm + Adam, Heusenstamm

Foreword

This is a translation of "Autour du théorème de Mordell-Weil", a course given by J.-P. Serre at the Collège de France in 1980 and 1981.
These notes were originally written weekly by Michel Waldschmidt and have been reproduced by Publications Mathématiques de l'Université de Paris VI, by photocopying the handwritten manuscript.
The present translation follows roughly the French text, with many modifications and rearrangements. We have not tried to give a detailed account of the new results due to Faltings, Raynaud, Gross-Zagier ...; we have just mentioned them in notes at the appropriate places, and given bibliographical references.

Paris, Fall 1988

M. L. Brown
J.-P. Serre

CONTENTS

1. SUMMARY

The aim of this course is the study of rational and integral points on algebraic varieties, especially on curves or abelian varieties. Before the end of the last century only special cases had been considered. The first general results are found, around 1890, in the work of Hurwitz and Hilbert [HH] where they introduced the, nowadays natural, viewpoint of algebraic geometry: if X and X' are two birationally equivalent algebraic curves over $\mathbf{Q}$ in $\mathbf{P}_2$, their rational points correspond. Hence the importance of birational invariants, in particular the genus. They studied especially the case of genus zero:

Theorem. *A curve of genus zero is isomorphic to a conic. If it has a rational point over $\mathbf{Q}$, then it is isomorphic to $\mathbf{P}_1$, and thus has an infinite number of rational points over $\mathbf{Q}$.*

The principle of the proof is the following. As the genus is zero, the canonical divisor has degree -2. Changing the sign, one obtains a divisor of degree 2. The correspondence between divisors and morphisms to projective space provides a morphism $X \to \mathbf{P}_2$ whose image is of degree 2, hence is a conic.

Around 1901, Poincaré [P] took up this question again, apparently unaware of the work of Hurwitz and Hilbert. He gave a clearer treatment of the case $g = 0$, then in the case of genus 1 (elliptic curves) he showed that if a rational point on the curve is chosen as an origin, the rational points form a group. He also gave some descent arguments (in fact a 3-descent, although his successors used mostly 2-descents). The conjecture that the group of rational points on a curve of genus 1 is finitely generated may be attributed to him. In 1922, Mordell [M] proved this:

Theorem. *The group of rational points on an elliptic curve is finitely generated.*

In the same paper Mordell stated his famous conjecture [now proved by Faltings]:

Mordell's Conjecture. *A curve of genus ≥ 2 has only finitely many rational points.*

In his thesis (1928), Weil [W1] considered a curve of genus $g \geq 1$ over a number field K, with jacobian J. Weil's theorem concerns the group $J(K)$ of points of J rational over K:

Theorem. (Mordell-Weil.) *The group $J(K)$ is finitely generated.*

One of the main themes of this course is the proof of this theorem (with J replaced by an arbitrary abelian variety).

Here is a brief sketch of the contents of each section:

1.1. Heights.

For simplicity, let us assume that K is $\mathbf{Q}$. Let $\mathbf{P}_n$ be projective n-space and $P \in \mathbf{P}_n(\mathbf{Q})$ a rational point of $\mathbf{P}_n$. We may write P uniquely (up to sign) as

$$P = (x_0, \ldots, x_n),$$

where $x_0, \ldots, x_n$ are rational integers with no common factor. Define the height of P to be

$$H(P) = \sup |x_i| .$$

The number of points of $\mathbf{P}_n(\mathbf{Q})$ of height $\leq N$ is asymptotically

$$\frac{2^n}{\zeta(n+1)} N^{n+1}.$$

The logarithmic height of P is $h(P) = \log H(P)$ (which is approximately the number of digits required to write the coordinates of P). If $\phi : X \to \mathbf{P}_n$ is a morphism, we define on $X(\mathbf{Q})$: $H_\phi(x) = H(\phi(x))$, $h_\phi(x) = h(\phi(x))$. If X is a projective variety, one associates to $\phi : X \to \mathbf{P}_n$ the divisor class $c_\phi \in \operatorname{Pic}(X)$ of the inverse image of a hyperplane section of $\mathbf{P}_n$. Then h_ϕ depends only on c_ϕ up to $O(1)$ (where $O(1)$ denotes a bounded function). Let $c \in \operatorname{Pic}(X)$; there are two morphisms ϕ and ψ such that $c = c_\phi - c_\psi$. We put $h_c = h_\phi - h_\psi$, which is defined up to $O(1)$. Then there are properties of the type $h_{c+c'} = h_c + h_{c'} + O(1)$. If moreover X is an abelian variety, there is the Néron normalisation of the height h_c: to every $c \in \operatorname{Pic}(X)$ is associated a well-defined function $\tilde{h}_c$. For simplicity, suppose that c is symmetric, that is to say $[-1]c = c$, where $[-1]$ is the morphism $x \mapsto -x$ on X. The Néron function is characterised by

$$\tilde{h}_c(nx) = n^2 \tilde{h}_c(x), \quad \text{for} \quad x \in X(\mathbf{Q}) \quad \text{and} \quad n \in \mathbf{Z}.$$

(In fact, this holds over $X(\bar{\mathbf{Q}})$.) This function $\tilde{h}_c$ is a quadratic form. There is also a decomposition of $\tilde{h}_c$ into local components.

1.2. The Mordell-Weil theorem and Mordell's conjecture.

The proof of the Mordell-Weil theorem divides into two steps.

The weak Mordell-Weil Theorem. Let A be an abelian variety defined over $\mathbf{Q}$, $\Gamma = A(\mathbf{Q})$ the group of rational points on A, and n an integer ≥ 1. *Then* $\Gamma/n\Gamma$ *is a finite group.*

The proof goes as follows. Suppose that $Y \to X$ is a finite étale covering of a projective variety X, where Y and X are defined over $\mathbf{Q}$. Let m be the degree of the covering. For each $x \in X(\mathbf{Q})$ we choose a lifting $y \in Y$. Let K_y be the field of rationality of y. Evidently, $[K_y : \mathbf{Q}] \leq m$. Further, one can show (Chevalley-Weil [CW]) that K_y is unramified outside a finite set of prime numbers $p_1, \ldots, p_h$, (depending only on X, Y and the covering map). But, by a theorem of Hermite, there are up to isomorphism only finitely many number fields of given discriminant and given degree. Thus the fields K_y are finite in number, and there is a number field K for which all rational points of X can be lifted to points of $Y(K)$ (that is to say, the image of $Y(K)$ contains $X(\mathbf{Q})$). One applies this to $Y = X = A$ and to the morphism $A \to A$ of multiplication by n. The finiteness of $\Gamma/n\Gamma$ follows easily.

The second part of the proof of the Mordell-Weil theorem is a descent argument using heights. Select an integer $n \geq 2$. Let x_i be representatives of the classes of Γ modulo $n\Gamma$. One shows that there is an integer r such that the points of Γ satisfying $h(x) \leq r$ generate Γ: choosing r large, if $h(x) > r$ one has that $x = ny + x_i$ with $y \in \Gamma$ and $h(y) < h(x)/2$; the proof then proceeds by induction.

As for Mordell's conjecture, it has been the subject of work of (a) Chabauty [Ch], (b) Demjanenko [D], and Manin [Ma1], (c) Mumford [Mu1], [and (d) Faltings [F]].

(a) Let X be an algebraic curve of genus ≥ 1 and J its jacobian. Suppose that the genus of X is strictly larger than the rank of $J(\mathbf{Q})$. Then $X(\mathbf{Q})$ is finite (Chabauty [Ch]).

(b) Let $x_0 \in X$. Let A be an abelian variety over $\mathbf{Q}$. Suppose that

$$\text{rank Hom}(X, A) > \text{rank } A(\mathbf{Q}),$$

where $\text{Hom}(X, A)$ denotes the group of morphisms from X to A which send x_0 to the neutral element of A. Then Demjanenko and Manin showed that $X(\mathbf{Q})$ is finite. This result has applications to modular curves.

(c) Let X be a curve of genus ≥ 2 over $\mathbf{Q}$. Suppose that it has an infinite number of rational points $P_1, P_2, \ldots$ ordered by increasing height. Then

Mumford proved,

$$h(P_n) \geq c_1 c_2^n, \text{ with } c_1 > 0, c_2 > 1.$$

[(d) Finally, Faltings proved Mordell's conjecture: if X is a curve of genus ≥ 2 over $\mathbf{Q}$, then $X(\mathbf{Q})$ is finite.]

1.3. Integral points on algebraic curves. Siegel's theorem.

Let X be an affine curve over $\mathbf{Q}$. and $\bar{X}$ its projective completion. The algebra Λ of regular functions on X is of finite type over $\mathbf{Q}$; let $x_1, \ldots, x_n \in \Lambda$ be a set of generators. A subset S of $X(\mathbf{Q})$ is *quasi-integral* if there is $\lambda \neq 0$ such that $x_i(S) \in (1/\lambda)\mathbf{Z}(1 \leq i \leq n)$; in other words, the $x_i(S)$ have bounded denominators. One checks that this condition is independent of the choice of generators.

Theorem (Siegel [Si]). *If $g \geq 1$, or if $g = 0$ and $\bar{X} - X$ has at least 3 elements, then all quasi-integral subsets are finite.*

The hypotheses of this theorem are necessary: if $g = 0$, the case of just one point at infinity corresponds to the affine line $\mathbf{A}_1$ (which has an infinite number of $\mathbf{Z}$-integral points), and the case where there are two points at infinity corresponds to the Pell-Fermat equation $X^2 - DY^2 = 1$.

Siegel's proof uses on the one hand, the weak Mordell-Weil theorem, and on the other hand a theorem on Diophantine approximation (Thue-Siegel-Roth theorem).

The idea consists of approximating the slope of an asymptote by rational numbers. For example, if the curve $X^3 - 3Y^3 = \text{const.}$ had an infinite number of integral points, one would immediately obtain better approximations of $3^{1/3}$ than are permitted by Roth's theorem. In the general case, one uses a theorem of bad approximation on abelian varieties:

Theorem. *Let A be an abelian variety defined over $\mathbf{Q}$, and let $x_1, x_2, \ldots$ be a sequence of distinct points of $A(\bar{\mathbf{Q}})$ converging to a point $x \in A(\mathbf{Q})$. Then*

$$\lim_{n \to \infty} \frac{\log 1/d(x, x_n)}{h(x_n)} = 0.$$

The proof uses Roth's theorem combined with the Mordell-Weil theorem.

The application to curves can be made in the following way:

Theorem. *Let X be a curve of genus ≥ 1 over $\mathbf{Q}$, f a non-constant rational function on X, and $x_1, x_2, \ldots$ an infinite number of distinct rational points. Put $f(x_i) = a_i/b_i, i = 1, 2 \ldots$, where, for $i \geq 1$, a_i and b_i are coprime integers. Then,*

$$\frac{\log |a_i|}{\log |b_i|} \to 1, \quad \text{as} \quad i \to \infty$$

For the proof, one embeds X in its jacobian and applies the theorem above.

1.4. Baker's Method.

For this method, the fundamental case is $g = 0$ with 3 points at infinity. The affine algebra of $\mathbf{P}_1 - \{0, 1, \infty\}$ is generated by 4 elements x, y, z, t with the relations

$$xy = 1, zt = 1, x + z = 1.$$

Therefore, associated to an integral point over K is a pair of units x, z of K with $x + z = 1$. More generally, the search for quasi-integral points reduces to the determination of pairs (x, z) of units of K satisfying some equation

$$Ax + Bz = C.$$

Expressing x, z in terms of a base of the group of units, one then uses Baker's theorem on lower bounds for linear forms in logarithms and obtains a finiteness result.

Other cases can be reduced to this by means of coverings; for example curves of genus 1 or curves of genus 2 with a pair of points invariant by the involution removed. Moreover the proof is *effective* (whereas Siegel's is not).

1.5. Hilbert's Irreducibility Theorem. Sieves.

Let C be an irreducible algebraic curve over $\mathbf{Q}$, and let $C \to \mathbf{P}_1$ be a morphism of degree $n \geq 2$. Then "few" rational points of $\mathbf{P}_1$ lift to rational points of C, and indeed, "most" points give extensions of degree n.

Equivalently, if X is the variable of $\mathbf{P}_1$ and $F(T, X) = 0$ is the equation of C, of degree n in T and irreducible over $\mathbf{Q}(X)$, there are "many" $x \in \mathbf{Q}$ for which $F(T, x)$ is irreducible over $\mathbf{Q}$.

The *rational* points of $\mathbf{P}_1$ with $H(P) \leq N$ which lift to rational points on C may be counted. Out of a total number $\sim cN^2$ rational points of height $\leq N$, the number of those which lift is

$$\begin{aligned}
&\ll N^{2/n} \ll N, && \text{if} \quad g = 0 \\
&\ll (\log N)^{\rho}, && \text{if} \quad g = 1 \quad \text{(Néron)} \\
&\ll \log\log N, && \text{if} \quad g \geq 2 \quad \text{(Mumford)} \\
&[\ll 1, && \text{if} \quad g \geq 2 \quad \text{(Faltings).]}
\end{aligned}$$

Out of a total of cN *integral* points on the affine line $\mathbf{A}_1$ with $H \leq N$, the number of those which lift is

$$\begin{aligned}
O(1) &\quad \text{if} \quad g \geq 1, \quad \text{or if} \quad g = 0 \quad \text{with } \geq 3 \text{ points at infinity} \\
O(\log N) &\quad \text{if} \quad g = 0 \quad \text{with 2 points at infinity} \\
O(\sqrt{N}) &\quad \text{if} \quad g = 0 \quad \text{with 1 point at infinity}
\end{aligned}$$

Hilbert's theorem can be stated for several variables: if $X \rightarrow \mathbf{P}_n$ is a morphism of degree ≥ 2, "few" rational points on $\mathbf{P}_n$ lift to $X(\mathbf{Q})$. For instance:

Theorem. *The number of integral points of $\mathbf{A}_n$ of height $\leq N$ which lift to $X(\mathbf{Q})$ is $\ll N^{n-1/2}(\log N)^{\gamma}$, for some γ.*

This result is due to S.D. Cohen [Co], with $\gamma = 1$.

Here is the principle of the proof. For simplicity, we consider the special case $z^2 = \phi(x_1, \ldots, x_n)$, where $\phi \in \mathbf{Z}[x_1, \ldots, x_n]$ has no square factors of degree ≥ 1. One wants to bound the number of points of

$$S_N = \{x_i \in \mathbf{Z}^n; |x_i| \leq N; f(x) \quad \text{is a square}\}.$$

As $z^2 - \phi$ is absolutely irreducible modulo p, for almost all p, a theorem of Lang-Weil shows that the number of points on the variety modulo p is $p^n + O(p^{n-1/2})$. As z and $-z$ give the same point, this shows that the reduction of S_N modulo p has $\ll (1/2 + \epsilon)p^n$ number of points. Since we have then excluded about half of the residue classes modulo p, we may apply the large sieve. In this particular case, it gives:

$$|S_N| \ll N^{n-1/2},$$

as desired.

Hilbert's theorem may be applied to construct field extensions of $\mathbf{Q}$ with given Galois group, such as S_n or A_n [or sporadic simple groups].

2. HEIGHTS

References: [L2] and [L6].

2.1. The product formula.

Let K be a field and M_K a family of absolute values $x \mapsto |x|_v$ (where $|x|_v$ is a real number ≥ 0) satisfying the following three properties:

1. All $v \in M_K$, except for a finite number, are ultrametric:

$$|x|_v = c^{-v(x)},$$

where v is a real valuation on K (always a discrete valuation in practice) and $c > 1$. The others (finite in number) are archimedean absolute values induced by embeddings $K \to \mathbf{C}$.

2. For all $x \in K^*$, one has $|x|_v = 1$ for all except finitely many v.

We suppose given a family λ_v of numbers > 0 and we put $||x||_v = |x|_v^{\lambda_v}$.

3. For all $x \in K^*$, one has (product formula):

$$\prod_{v \in M_K} ||x||_v = 1.$$

We then say that K is equipped with a product formula.

Examples. a. $K = \mathbf{Q}$. $M_{\mathbf{Q}} = P \cup \{\infty\}$ where P is the set of prime numbers, $\lambda_v = 1$ for all $v \in M_{\mathbf{Q}}$. For $x \in \mathbf{Q}^*$, we write

$$x = \pm \prod_{p \in P} p^{\nu_p(x)}$$

and put

$$||x||_\infty = |x| = \prod_{p \in P} p^{\nu_p(x)} (\text{usual absolute value of } x)$$

$$||x||_p = |x|_p = p^{-\nu_p(x)}, \quad \text{for} \quad p \in P.$$

b. Let V be a projective normal k-variety, K its function field, M_V the set of irreducible divisors W of V (i.e. irreducible subvarieties of codimension 1). As V is normal, for all $W \in M_V$ the local ring of W is a discrete valuation ring. We write $v_W : K^* \to \mathbf{Z}$, where $v_W(f)$ is the order of f on W , and

$$|f|_W = c^{-v_W(f)}, \qquad (f \in K^*).$$

As each $f \in K^*$ has only a finite number of zeros and poles, we have $v_W(f) = 0$ except for a finite number of $W \in M_V$.

Let C be a curve on V not intersecting the singularities of V (as V is normal, its singularities have codimension at least 2; one obtains such a curve by taking, for example, a generic section by a linear variety or by a hypersurface of sufficiently large degree). Let $C.W$ be the intersection multiplicity of C and W (this number is defined first when W does not contain C, and one checks that it depends only on the class of W; this then allows $C.W$ to be defined when $W \supset C$).

We assume that $C.W > 0$ *for all* $W \in M_V$. This is the case, for example, when C is a section of V by a linear variety of the right dimension. We then put $\lambda_W = C.W$.

For $f \in K^*$, (f) is linearly equivalent to zero, thus $C.(f) = 0$ (if C is not a pole of f , then $C.(f)$ is the degree of the divisor on C given by the restriction of f to C, that is to say the number of zeros minus the number of poles). As

$$(f) = \sum_W v_W(f) W,$$

we get

$$\sum_W \lambda_W v_W(f) = 0 \quad \text{for all } f \in K^*.$$

We may thus obtain different product formulae for the same field K (for example, take a quadric surface in 3-dimensional space with different choices of C).

c. *Finite field extensions.*

Let K be a field with a product formula, and let L be a finite extension of K of degree $[L : K] = d$. Each $v \in M_K$ is known to extend to a finite number of absolute values w of L; call M_L the set of such extensions. Using M_L, a product formula for L is obtained in the following way:

Write K_v for the completion of K at v. Then $L_v = L \otimes_K K_v$ is a semi-local algebra of degree d over K_v; hence it is a product of local algebras

$$L_v = \prod_{w|v} \tilde{L}_w,$$

where the residue field L_w of $\tilde{L}_w$ is the completion of L at w. (If L/K is separable - which is the main case for applications - then $\tilde{L}_w = L_w$.) Let us write

$$\tilde{d}_w = [\tilde{L}_w : K_v] \text{ and } d_w = [L_w : K_v].$$

For $x \in L$, put:

$$||x||_w = ||N_{L_w/K_v}(x)||_v^{\frac{\tilde{d}_w}{d_w} \cdot \frac{1}{d}}$$

(In the separable case this becomes simply

$$||x||_w = ||N_{L_w/K_v}(x)||_v^{1/d} \quad .)$$

Then for all $x \in L^*$ we have

(1) $$\prod_{w|v} ||x||_w = ||N_{L/K}(x)||^{1/d}.$$

If $x \in K^*$, we get

(2) $$\prod_{w|v} ||x||_w = ||x||_v.$$

Formula (1) implies that $\prod_{w \in M_L} ||x||_w = 1$: we have a product formula for L.

Proof of (1).

Denote by $\tilde{x}_w$ the image of $x \in L^*$ in $\tilde{L}_w$. On taking the norm from $L \otimes_K K_v$ to K_v, one gets

$$N_{L/K}(x) = \prod_{w|v} N_{\tilde{L}_w/K_v}(\tilde{x}_w).$$

Let x_w be the image of $\tilde{x}_w$ by the projection $\tilde{L}_w \to L_w$. We have a Jordan-Hölder series of $\tilde{L}_w$ (considered as a module over itself) of length $\tilde{d}_w/d_w$, with factor modules isomorphic to L_w. We deduce

$$N_{\tilde{L}_w/K_v}(\tilde{x}_w) = (N_{L_w/K_v}(x_w))^{\tilde{d}_w/d_w},$$

whence (1). It is clear that (1) $\Rightarrow$ (2).

Number fields.

Let K be a number field of degree d over $\mathbf{Q}$. Let v be a place of K. We have defined above the normalised absolute value $||x||_v$.

Another way to define it is the following: for $x \in K$, the homothety $y \mapsto xy$ transforms the additive Haar measure on K_v to a multiple of itself and the multiplier $|x|_{v,\mathrm{Haar}}$ is related to $||x||_v$ by:

$$||x||_v = |x|_{v,\mathrm{Haar}}^{1/d}.$$

If v is associated to a real embedding $K \to \mathbf{R}$, then

$$||x||_v = |x|^{1/d}.$$

If v is associated to a complex embedding $K \to \mathbf{C}$, then

$$||x||_v = |x|^{2/d}.$$

We have:

$$||x_1 + \ldots + x_N||_v \leq \begin{cases} \sup_{1 \leq i \leq N} ||x_i||_v & \text{if } v \text{ is ultrametric,} \\ ||N||_v \sup ||x_i||_v, & \text{if } v \text{ is archimedean.} \end{cases}$$

2.2. Heights on $\mathbf{P}_m(K)$.

Let K be a field with a product formula. By §2.1 example (c), every subfield L of an algebraic closure $\bar{K}$ of K, and of finite type over K, has a product formula.

Let $x \in \mathbf{P}_n(K)$, $x = (x_0, \ldots, x_n)$, with $x_i \in K$ not all zero. We put

$$H(x_0, \ldots, x_n) = \prod_{v \in M_K} \sup_{0 \leq i \leq n} ||x_i||_v.$$

Almost all terms in the product are equal to 1. Further

$$H(\lambda x_0, \ldots, \lambda x_n) = (\prod_v ||\lambda||_v) H(x_0, \ldots, x_n),$$

$$= H(x_0, \ldots, x_n),$$

for all $\lambda \in K^*$; this defines H on $\mathbf{P}_n(K)$. Finally, we have $H(x) \geq 1$ for all $x \in \mathbf{P}_n(K)$, for if $x_0 \neq 0$, for example, then $H(x) \geq \prod_v ||x_0||_v \geq 1$.

We put $h(x) = \log H(x)$. Then, $h(x) \geq 0$. If L is a finite extension of K and $x \in \mathbf{P}_n(L)$ we define

$$H_L(x) = \prod_{w \in M_L} \sup_i ||x_i||_w.$$

Then

$$H_L(x) = H_K(x), \quad \text{for} \quad x \in \mathbf{P}_n(K).$$

Indeed, for each $v \in M_K$ let i_v be such that

$$||x_{i_v}||_v \geq ||x_j||_v \quad \text{for all } j.$$

Then, $||x_{i_v}||_w \geq ||x_j||_w$ for all j and all $w|v$. Therefore

$$\prod_{w|v} \sup_j ||x_j||_w = \prod_{w|v} ||x_{i_v}||_w = ||x_{i_v}||_v.$$

We have thus defined a height $H : \mathbf{P}_n(\bar{K}) \rightarrow \mathbf{R}$ with values ≥ 1, and a logarithmic height $h : \mathbf{P}_n(\bar{K}) \rightarrow \mathbf{R}_+$ with values ≥ 0.

Examples

a. $K = \mathbf{Q}$.

For $x \in \mathbf{P}_n(\mathbf{Q})$, we may write $x = (x_0, \ldots, x_n)$ where $x_i \in \mathbf{Z}$ have greatest common divisor 1. For all primes p one has $v_p(x_i) \geq 0$, and $v_p(x_i) = 0$ for some i. Thus, $\inf_i v_p(x_i) = 0$ and $\sup_i ||x_i||_p = 1$. Therefore, $H(x) = \sup_i |x_i|$.

b. *Number fields.*

For a number field, there is an analogous formula: Let K be a number field, O_K the ring of integers of K, and $M_{K,\infty}$ the set of archimedean places of M_K. To $x = (x_0, \ldots, x_n) \in \mathbf{P}_n(K)$, let us associate the ideal $\underline{a} = x_0 O_K + \ldots + x_n O_K$; we have:

$$H(x) = 1/N(\underline{a})^{1/d} \prod_{v \in M_{K,\infty}} \sup_i ||x_i||_v.$$

c. *Function fields.*

Let $K = k(V)$ be the function field of a projective normal k-variety V. An element $x = (x_0, \ldots, x_n) \in \mathbf{P}_n(K)$ can be considered as a rational map of V to $\mathbf{P}_n$:

$$x : V \cdots \rightarrow \mathbf{P}_n,$$

(the set Δ of points of V where x is not defined has codimension ≥ 2). Thus x is a morphism of $V - \Delta$ to $\mathbf{P}_n$ with codim $\Delta \geq 2$.

Let H be a general hyperplane of $\mathbf{P}_n$ (not containing the image of x) and let $x^*(H)$ be the inverse image of H; $x^*(H)$ is a divisor on $V - \Delta$, thus its Zariski closure is a divisor on V.

Let C be the curve chosen to define the product formula on K, cf. §2.1, example b).

Formula.
$$h(x) = C.x^*(H)$$

It is enough to prove the formula in the particular case where $x_i \neq 0$ for all i. Denote by H_i the coordinate hyperplanes in $\mathbf{P}_n$, $D_i = x^*(H_i)$, and $D_i - D_j = (x_i/x_j)$. For W an irreducible divisor on V, denote by $v_W(D_i)$ the coefficient of W in D_i. We have

$$v_W(D_i) - v_W(D_j) = v_W(x_i) - v_W(x_j).$$

Taking $j = 0$:

$$\inf_i v_W(D_i) - v_W(D_0) = \inf_i v_W(x_i) - v_W(x_0).$$

As $D_i \geq 0$, we have $v_W(D_i) \geq 0$; as the H_i have no common points, $\inf_i v_W(D_i) = 0$. Therefore

$$-(W.C)v_W(D_0) = (W.C)\inf_i v_W(x_i) - (W.C)v_W(x_0).$$

But

$$D_0 = \sum_W W.v_W(D_0),$$

and

$$C.D_0 = \sum_W (C.W)v_W(D_0).$$

On the other hand

$$\sum_W (W.C)v_W(x_0) = C.(x_0) = 0,$$

whence

$$-C.D_0 = \sum_W (W.C)\inf_i v_W(x_i)$$

$$= -h(x),$$

which proves the formula.

2.3. Properties of heights.

1. If x has coordinates x_i and y has coordinates y_j, let $x \otimes y$ be the point with coordinates $x_i.y_j$. Then:

$$H(x \otimes y) = H(x).H(y).$$

2. For $x = (x_0, \ldots, x_n)$ and $m \geq 1$, let $x^{(m)}$ be the point whose projective coordinates are all the monomials of degree m in the x_i, and $x^m = (x_o^m, \ldots, x_n^m)$. Then:

$$H(x^{(m)}) = H(x^m) = H(x)^m.$$

3. *Formula for changing coordinates.* Let $\phi_0, \ldots, \phi_r$ be homogeneous polynomials of degree m in $X_0, \ldots, X_n$ with coefficients in K. If $x = (x_0, \ldots, x_n) \in \mathbf{P}_n(K)$, and if the $\phi_i(x)$ are not all zero, define $\phi(x) = (\phi_0(x), \ldots, \phi_r(x)) \in \mathbf{P}_r(K)$. Then:

$$H(\phi(x)) \leq ||N||_\infty H(c) H(x)^m$$

where c and N are defined in the following way. Each ϕ_i is a linear combination of monomials in $X_0, \ldots, X_n$ of degree m; c is the sequence of coefficients of all the ϕ_i considered as a point in projective space; N is the maximum number of monomials appearing in any one of the ϕ_i; finally

$$||N||_\infty = \prod_{v \in M_\infty} ||N||_v$$

(if K is a number field with the usual product formula, $||N||_\infty = N$).

Corollary. *For fixed* ϕ, *we have*

$$H(\phi(x)) = O((H(x))^m) \quad \textit{for} \quad x \in \mathbf{P}_n(\bar{K}).$$

In particular if $\phi \in \mathrm{PGL}_{n+1}(K)$ is an automorphism of $\mathbf{P}_n$, we have

$$h(\phi(x)) = h(x) + O(1),$$

i.e. the difference is a bounded function on $\mathbf{P}_n(\bar{K})$. If one neglects bounded functions, the logarithmic height is independent of the chosen coordinates on $\mathbf{P}_n$.

4. Let $F(X_0, \ldots, X_n)$ be a homogeneous polynomial of degree m in which the coefficient of x_n^m is non-zero, that is to say $F(0, \ldots, 0, 1) \neq 0$. For $x = (x_0, \ldots, x_n)$, let $x' = (x_0, \ldots, x_{n-1})$.

Proposition. *There is a constant C depending only on F such that if $x \in \mathbf{P}_n(\bar{K})$ satisfies $F(x) = 0$, then*

$$h(x) - C \leq h(x') \leq h(x).$$

(If $F(x) = 0$ then $x \neq (0, \ldots 0, 1)$ thus x' defines a point of $\mathbf{P}_{n-1}(\bar{K})$.)

Proof.

Let x'' be the point whose coordinates are all the monomials of degree m in $x_0, \ldots, x_n$, with the exception of x_n^m. As $F(x) = 0$, x_n^m is a linear combination of other monomials, thus the map $x'' \mapsto x^{(m)}$ is linear, and

$$h(x^{(m)}) \leq h(x'') + O(1).$$

The point $x^{(m-1)} \otimes x'$ has as coordinates the monomials in $x_0, \ldots, x_n$ of degree m containing at least one variable different from x_n. Up to repetitions of the coordinates this is the point x'', hence:

$$h(x'') = h(x^{(m-1)}) + h(x').$$

As $h(x^{(m)}) = mh(x)$ and $h(x^{(m-1)}) = (m-1)h(x)$, we have the result.

An explicit bound for the constant C can be given namely:

$$C \leq \log ||N||_\infty + h(c),$$

where N is the number of monomials occuring in F apart from x_n^m, and c is the sequence of coefficients of F.

Geometric interpretation.

The map $x \mapsto x'$ defines a morphism of $\mathbf{P}_n - \{e\}$ to $\mathbf{P}_{n-1}$, with $e = (0, \ldots, 0, 1)$; this is the central projection from the point e.

More generally, if Λ is a linear subvariety of $\mathbf{P}_n$ of dimension λ, one has (on selecting some coordinates) a projection $\pi : \mathbf{P}_n - \Lambda \longrightarrow \mathbf{P}_{n-\lambda-1}$.

Proposition. *If $X \subset \mathbf{P}_n$ is a closed projective subvariety not intersecting Λ, then*

$$h(\pi(x)) = h(x) + O(1), \qquad \textit{for all } x \in X(\bar{K}),$$

where the $O(1)$ depends only on X.

Proof.

Induction on λ. The case $\lambda = 1$ is the previous proposition. It is only necessary to check that if X is a closed sub-variety of $\mathbf{P}_n$, not meeting Λ, then the image of X is closed in $\mathbf{P}_{n-\lambda-1}$. This follows from the fact that a projective variety is complete, thus its image by a morphism is closed.

We may also see this more concretely. It suffices to consider a projection whose centre is a point ($\lambda = 0$). Let F, F_α be the equations defining X, with $F(0,\ldots,0,1) \neq 0$. One wishes to find the equations $G_\beta(x_0,\ldots,x_{n-1}) = 0$ of the projection, that is to say, to eliminate x_n from the equations F_α. Let us write

$$F(x) = x_n^m + a_1(x_0,\ldots,x_{n-1})x_n^{m-1} + \ldots,$$

and we factorise it formally as:

$$F(x) = \prod_{i=1}^{m}(x_n - \zeta_i(x_0,\ldots,x_{n-1})).$$

We introduce indeterminates t_α and put:

$$\prod_{i=1}^{m}\sum_{\alpha} t_\alpha F_\alpha(x_0,\ldots,x_{n-1},\zeta_i) = \sum_a t^a G_a(x_0,\ldots,x_{n-1}).$$

One checks that the G_a give the equations of the projection. (If there is a single F_α, the resultant of F and F_α is then obtained.)

Proposition. *Let $\phi_0,\ldots,\phi_r$ be homogeneous polynomials of degree m which are not simultaneously zero on $\bar{K}$. Then*

$$h(\phi(x)) = mh(x) + O(1), \quad \textit{for } x \in \mathbf{P}_n(\bar{K}).$$

Proof.

Let $i : \mathbf{P}_n \to \mathbf{P}_M$ be defined by $i(x) = x^{(m)}$, where $M+1$ is the number of monomials of degree m. There is a linear projection, defined outside the closed subvariety of its centre Δ, $\pi : \mathbf{P}_M - \Delta \to \mathbf{P}_r$ such that $\pi \circ i = \phi$.

$$\begin{array}{ccc} \mathbf{P}_n & \xrightarrow{i} & \mathbf{P}_M - \Delta \\ \downarrow{\scriptstyle\phi} & \swarrow{\scriptstyle\pi} & \\ \mathbf{P}_r & & \end{array}$$

Then $i(\mathbf{P}_n) \cap \Delta = \emptyset$. Thus on $i(\mathbf{P}_n)$ we have $h(\pi(y)) = h(y) + O(1)$. As $y = i(x) = x^{(m)}$ the result follows.

Exercise. (Alternative proof.) Use Hilbert's Nullstellensatz to show that there is an integer A for which x^A is the image of $x^{(A-m)} \otimes \phi(x)$ by a linear transformation. Hence

$$mh(x) \leq h(\phi(x)) + O(1),$$

and the reverse inequality is obvious.

2.4. Northcott's finiteness theorem.

Theorem ([No]). *Let n, d, X be integers ≥ 1. There are only finitely many points of $\mathbf{P}_n(\bar{\mathbf{Q}})$ of height $\leq X$ and of degree $\leq d$.*

For $x = (x_0, \ldots, x_n) \in \mathbf{P}_n(\bar{\mathbf{Q}})$, the degree of x is the degree of the field generated by the $x_i/x_j, 0 \leq i, j \leq n, x_j \neq 0$.

Corollary. *If K is a finite extension of $\mathbf{Q}$, there are only finitely many points of $\mathbf{P}_n(K)$ of height $\leq X$.*

Proof of the theorem.
a). $d = 1$. The points of $\mathbf{P}_n(\mathbf{Q})$ of bounded height are finite in number. Indeed, let us write $x \in \mathbf{P}_n(\mathbf{Q})$ as $x = (x_0, \ldots, x_n)$, $x_j \in \mathbf{Z}$ with greatest common divisor 1. Then $H(x) = \sup |x_i|$, whence the finiteness.
b). We reduce the case $d \geq 1$ to the case $d = 1$.

Let $x = (x_0, \ldots, x_n) \in \mathbf{P}_n(K)$, with $[K : \mathbf{Q}] = d$, $x_i \in K$. Let $\sigma_\alpha : K \to \bar{\mathbf{Q}}$, $(1 \leq \alpha \leq d)$ be the embeddings of K in $\bar{\mathbf{Q}}$, and $x^\alpha = \sigma_\alpha x = (x_0^\alpha, \ldots, x_n^\alpha)$. Thus $(x^1, \ldots, x^d) \in \mathbf{P}_n \times \ldots \times \mathbf{P}_n(\bar{\mathbf{Q}})$. The symmetric group S_d of degree d acts on $\mathbf{P}_n \times \ldots \times \mathbf{P}_n$. Let $\Sigma_{n,d} = (\mathbf{P}_n \times \ldots \times \mathbf{P}_n)/S_d$, and let $\pi : \mathbf{P}_n \times \ldots \times \mathbf{P}_n \to \Sigma_{n,d}$ be the natural morphism. The image of $\underline{x} = (x^1, ..., x^d)$ by π is invariant under conjugation, thus $\pi(\underline{x}) \in \Sigma_{n,d}(\mathbf{Q})$. There is a natural immersion $\epsilon : \Sigma_{n,d} \to \mathbf{P}_N$ (consider the homogeneous polynomials in $x_0^\alpha, \ldots, x_n^\alpha$ for each α, acted on by the symmetric group, and take invariant polynomials). Then

$$H(\epsilon\pi(\underline{x})) \leq C \prod_\alpha H(x^\alpha)^{d_\alpha} \leq CH(x)^D, \text{ for some } d_\alpha, C, D.$$

Using a), this shows that the $\epsilon\pi(\underline{x})$ are finite in number. Each fibre has at most $d!$ elements, whence the finiteness of the x's.

Exercise. Derive part (b) of this proof by introducing variables $T_0, \ldots, T_n$, and considering the polynomial

$$\prod_{\alpha=1}^{d} \sum_{i=0}^{n} T_i \sigma_\alpha(x_i) = \sum_a c_a(x) T^a.$$

Then $c_a(x) \in \mathbf{Q}$ and the point $c(x)$ with coordinates $c_a(x)$ is such that:

$$H(c(x)) \leq CH(x)^D.$$

Conclude as above.

2.5. Quantitative form of Northcott's theorem.

Theorem (Schanuel [Sch]). *Let N_X be the number of $x \in \mathbf{P}_{n-1}(K)$ with $H(x) \leq X$. Then*

$$N_X = \frac{h}{\zeta_K(n)} c_n X^{nd} + \begin{cases} O(X \log X), & \text{if } d = 1,\ n = 2, \\ O(X^{nd-1}), & \text{otherwise.} \end{cases}$$

where

$$c_n = \left(\frac{2^{r_1}(2\pi)^{r_2}}{\sqrt{\Delta}}\right)^n \frac{R}{w} n^r.$$

(As usual, r_1 is the number of real places, r_2 the number of complex places, $r = r_1 + r_2 - 1$ the rank of the group of units, $d = [K : \mathbf{Q}]$, w the number of roots of unity in K, R the regulator, Δ the absolute value of the discriminant, and h is the class number).

Remark. The error term is probably not best possible for $(n, d) \neq (2, 1)$.

Sketch of proof.

Let $x = (x_0, \ldots, x_{n-1}) \in \mathbf{P}_{n-1}(K)$. Let $\underline{a}_x$ be the ideal generated by the x_i. For $\underline{a}$ an ideal of K, let $N_X(\underline{a})$ be the number of points of height $\leq X$ with class$(\underline{a}_x) =$ class$(\underline{a})$. Schanuel proves the more precise estimate:

$$N_X(\underline{a}) = 1/\zeta_K(n) c_n X^{nd} + O(-).$$

Recall that if $\underline{a}_x = \underline{a}$, we have

$$H(x) = N(\underline{a})^{-1/d} \prod_{v \in M_{K,\infty}} \sup ||x_i||_v = N(\underline{a})^{-1/d} H_\infty(x).$$

We are thus led to study the n-tuples $(x_1, \ldots, x_n)$, with $x_i \in \underline{a}$, such that a) $H_\infty(x) \leq X N(\underline{a})^{1/d}$, and b) the x_i generate $\underline{a}$.

We define an equivalence relation on $\underline{a} \times \ldots \times \underline{a} = \underline{a}^n$:

$$(x_1, \ldots, x_n) \sim (y_1, \ldots, y_n) \quad \text{if there is a unit } \epsilon \text{ of } K \text{ with } y_i = \epsilon x_i \text{ for all } i.$$

Let $N'_X(\underline{a})$ be the number of $(x_1, \ldots, x_n) \in \underline{a}^n$, modulo this equivalence relation, with $H_\infty(x) \leq X N(\underline{a})^{1/d}$.

Theorem. $N'_X(\underline{a}) = c_n X^{nd} + O_{K,n}(X^{nd-1})$.

Since $N'_X(\underline{a})$ depends only on the class of $\underline{a}$, which takes only finitely many values, the O-constant can be chosen to be independent of $\underline{a}$.

a. *Computation of* $N'_X(\underline{a})$.

Let $K_{\mathbf{R}} = K \otimes \mathbf{R}$ (product of the archimedian completions of K). In $K_{\mathbf{R}} \times \ldots \times K_{\mathbf{R}}$ (n times) we count the points of the lattice $\underline{a} \times \ldots \times \underline{a}$. For this we take a fundamental domain Σ of $K_{\mathbf{R}}$ modulo the units. For each $v \in M_{K,\infty}$ let $i(v)$ be such that $||x_{i(v)}||_v = \sup_i ||x_i||_v$. We then have a point $(x_{i(v)}) \in \prod K_v$. Define a domain D_X of $K^n_{\mathbf{R}}$ by the 3 conditions:

(1) $(x_{i(v)}) \in \Sigma$,

(2) $H_\infty(x) \leq XN(\underline{a})^{1/d}$,

(3) $||x_j||_v \leq ||x_{i(v)}||_v$ for all j and all v.

It remains to count the points of the lattice in D_X. Notice that $H_\infty(\lambda x) = \lambda H_\infty(x)$, and $D_X = XD_1$. It is easily shown that the frontier $F = \partial D_1$ of the compact set D_1 is contained in a finite number of images of $[0,1]^{nd-1}$ by morphisms ϕ_i of class C^1 and therefore is $(nd-1)$-dimensional, in Minkowski's sense.

Given a compact subset D_1 of $\mathbf{R}^N$ whose frontier has the above property, the number of points of $\mathbf{Z}^N$ which belongs to XD_1 is $\mathrm{vol}(D_1)X^N + O(X^{N-1})$. [The idea of the proof essentially derives from Gauss, and consists of counting the number of translates of a fundamental domain for the lattice which intersect XD_1 and then showing that the number of translates which intersect the boundary XF is $O(X^{nd-1})$. This last estimate is derived by dissecting $[0,1]^{nd-1}$ into $[X^{1/nd}]^{nd-1}$ small cubes whose images under the maps ϕ_i have diameter bounded independently of X, by the C_1 property.] The estimate for $N'_X(\underline{a})$ follows.

b. *Computation of* $N_X(\underline{a})$.

Let $(x_1, \ldots, x_n) \in \underline{a}^n$ with $H_\infty(x) \leq XN(\underline{a})^{1/d}$. The ideal generated by $x_1, \ldots, x_n$ is of the form $\underline{a}b$ with b an integer. Hence

$$N'_X(\underline{a}) = \sum_b N_{X(Nb)^{-1/d}}(\underline{a}b).$$

Using the Möbius formula, we get

$$N_X(\underline{a}) = \sum_b \mu(b) N'_{X(Nb)^{-1/d}}(\underline{a}b),$$

$$\sim \sum_b \mu(b) c_n X^{nd} (Nb)^{-n}$$

$$\sim c_n X^{nd} \sum_b \mu(b)(Nb)^{-n}.$$

Finally, Schanuel's theorem follows from

$$\frac{1}{\zeta_K(s)} = \sum_b \mu(b)(Nb)^{-s}.$$

The function field analogue.

Let C be an absolutely irreducible projective curve over $\mathbf{F}_q$ of genus g and let K be its function field. If $x = (x_0, \ldots, x_{n-1}) \in \mathbf{P}_{n-1}(K)$, we have $h(x) = \text{degree}(x^*H)$, where x^*H is the inverse image by $x : C \to \mathbf{P}_{n-1}$ of a hyperplane H of $\mathbf{P}_{n-1}$ not containing the image of C. One can prove:

Theorem. *The number of $x \in \mathbf{P}_{n-1}(K)$ with $h(x) = d$ is*

$$\frac{Aq^{n(d+1-g)}}{(q-1)\zeta_K(n)} + O(q^{n(d+1-g)-1}),$$

where A is the number of $\mathbf{F}_q$-rational points on the jacobian of C (= number of divisor classes of degree 0 of K).

2.6. Height associated to a morphism $\phi : X \to \mathbf{P}_n$.

Let K be a field equipped with a product formula, $\bar{K}$ an algebraic closure of K , and X an algebraic variety over K. Let $\phi : X \to \mathbf{P}_n$ be a morphism. For $x \in X(\bar{K})$, we put

$$H_\phi(x) = H(\phi(x)),$$

$$h_\phi(x) = \log H_\phi(x) = h(\phi(x)).$$

Theorem. *Let $\phi : X \to \mathbf{P}_n$ be an isomorphism of X onto a locally closed subvariety of $\mathbf{P}_n$. Let $\psi : X \to \mathbf{P}_m$ be any morphism. There are real numbers $c_1 > 0$ and c_2 such that*

$$h_\psi \leq c_1 h_\phi + c_2 \quad \text{on} \quad X(\bar{K}).$$

In other words, $h_\psi \leq O(1 + h_\phi)$.

Proof.

We view X as embedded in $\mathbf{P}_n$ (by ϕ) with coordinates $x_0, \ldots, x_n$.

a) Special case: there are homogeneous polynomials $\psi_0, \ldots, \psi_m$ of the same degree, not simultaneously zero on X, such that

$$\psi(x) = (\psi_0(x), \ldots, \psi_n(x)).$$

Let d be the common degree of the ψ_i. Then

$$H(\psi(x)) \leq CH(x)^d.$$

b) General case. We cover X by a finite number of open subvarieties X_i such that the pairs $(X_i, \psi|_{X_i})$ are special.

By the quasi-compactness of algebraic varieties it suffices to do this locally. Let $x \in X$. We write

$$\psi(x) = (\lambda_0, \ldots, \lambda_n),$$

with, let's say, $\lambda_0 = 1$. Then in a neighbourhood of x

$$\psi = (1, f_1, \ldots, f_m),$$

where the f_i belong to the local ring of x :

$$f_i = \psi_i / \psi_0 \,,$$

where the ψ_i are homogeneous of the same degree and $\psi_0(x) = 1$. So $\psi = (\psi_0, \ldots, \psi_m)$ in a neighbourhood of x, and the result follows from a).

Remark. If ψ is also an immersion, we have

$$c_3 h_\phi + c_4 \leq h_\psi \leq c_1 h_\phi + c_2,$$

for some constants c_i, hence h_ϕ and h_ψ have roughly the same size.

2.7. The group Pic(X) (reference: [H]).

Let X be a variety over a field K. One defines Pic(X) to be the group of line bundles (= locally free sheaves of O_X-modules of rank 1) on X, the multiplication being the tensor product; one has $\mathrm{Pic}(X) \cong H^1(X, O_X^*)$ ([H], Ch.III, Ex.4.5). If $f : X \to Y$ is a morphism and c is a line bundle on Y, then f^*c , which is a line bundle on X, denotes the inverse image of c by f. This defines a homomorphism

$$f^* : \mathrm{Pic}(Y) \to \mathrm{Pic}(X).$$

Suppose now that X is a normal variety (in fact, normality is unnecessary here, [G, Ch.IV, 21.6]). A (Weil) divisor on X is a linear combination, with integral coefficients, of irreducible sub-varieties of codimension 1:

$$D = \sum_W n_W . W.$$

The support Supp D of D is $\bigcup_{n_W \neq 0} W$.

If f is a rational function on X, its divisor is the divisor of zeros minus the divisor of poles:

$$(f) = (f)_0 - (f)_\infty = \sum_W v_W(f).W.$$

The quotient of the group of divisors by the subgroup of divisors of functions is the group $\mathrm{Cl}(X)$ of divisor classes. One says that the divisor D is linearly equivalent to zero, written $D \sim 0$, if it is the divisor of a rational function, that is, D is in the zero class.

A line bundle on X has rational sections; if s is such a section one defines its divisor $(s) = (s)_0 - (s)_\infty$, and the class of this divisor is independent of the section chosen. For a normal variety, one then obtains an injection $\mathrm{Pic}(X) \to \mathrm{Cl}(X)$.

In the non-singular case (locally factorial suffices, [H, Ch. II, Corollary 6.16]), this gives an isomorphism

$$\mathrm{Pic}(X) \cong \mathrm{Cl}(X).$$

On $\mathbf{P}_n$ we have a natural line bundle, written c_1, which corresponds to the class of a hyperplane. For a morphism $\phi : X \to \mathbf{P}_n$, the inverse image $\phi^* c_1 = c_\phi$ defines an element of $\mathrm{Pic}(X)$ (if H is a hyperplane not containing $\phi(X)$, X being irreducible and smooth, c_ϕ is the class of $\phi^* H$). Hence one has a map $\phi \mapsto c_\phi$.

If for any $x \in X$ there is a global section of the line bundle E on X which does not vanish at x, then E is said to be is generated by its global sections. Let X be projective and E a line bundle on X. Then $\Gamma(X, E)$ is finite dimensional. Furthermore, if E is generated by its global sections, a morphism ϕ from X to $\mathbf{P}_n$ is obtained in the following way. Let $s_0, \ldots, s_n$ be a basis of $\Gamma(X, E)$. The morphism $\phi : X \to \mathbf{P}_n$, where $\phi(x) = (s_0(x), \ldots, s_n(x))$ for all $x \in X$, is well defined, and we have $c_\phi = E$.

Suppose that X is non-singular and projective. A necesssary and sufficient condition for $c \in \mathrm{Pic}(X)$ to be generated by its global sections is that if the divisor D has class c, then the complete linear system $|D|$ has no base points: that is to say, for all $x \in X$, there is D' linearly equivalent to D with $D' \geq 0$ and $x \notin \mathrm{Supp}(D')$.

An element $c \in \mathrm{Pic}(X)$, which is generated by its global sections, is *very ample* if the corresponding morphism $\phi : X \to \mathbf{P}_n$ is an immersion. We say that c is *ample* if there is an integer $m > 0$ such that mc is very ample, in which case mc is very ample for all sufficiently large m.

2.8. Heights and line bundles

Let K be a field with a product formula, V a projective algebraic variety over K, and $\mathrm{Pic}(V)$ the group of classes of line bundles on V.

Let H be the quotient of the vector space of real-valued functions on $V(K)$ by the space of bounded functions on $V(K)$. It will be convenient to choose representatives of elements of H and to write that formulae are true modulo $O(1)$, that is to say up to the addition of a bounded function.

Theorem. *There is a unique map $c \mapsto h_c$ of Pic(V) to H such that,*
1) $h_{c+c'} = h_c + h'_c + O(1)$ for all $c, c' \in \mathrm{Pic}(V)$.
2) If c corresponds to a morphism $\phi : V \to \mathbf{P}_n$, then $h_c = h_\phi + O(1)$.

We recall that the correspondence between $\phi : V \to \mathbf{P}_n$ and $c_\phi \in \mathrm{Pic}(V)$ is given by $\phi^* c_1 = c_\phi$ where c_1 is the standard line bundle on $\mathbf{P}_n$. On the other hand, $h_\phi(x) = h(\phi(x))$ where h is the logarithmic height on $\mathbf{P}_n$.

Proof.
Uniqueness. It suffices to prove the following result: *every $c \in \mathrm{Pic}(V)$ is of the form $c_\phi - c_\psi$ where ϕ, ψ are immersions.*

Let $\phi_1 : V \to \mathbf{P}_n$ be an immersion, and let $c_{\phi_1} = \phi_1^* c_1$.

a. If m is sufficiently large, $c' = c + mc_{\phi_1}$ is generated by its global sections. This is a general result on sheaves twisted by $\mathbf{O}(1)$.

b. As c' is generated by its global sections, it corresponds to a morphism $V \to \mathbf{P}_m$. We embed $\mathbf{P}_m \times \mathbf{P}_n$ in a $\mathbf{P}_N$ by the tensor product (Segre embedding). Then, $c' + c_{\phi_1}$ corresponds to an immersion $V \to \mathbf{P}_m \times \mathbf{P}_n \to \mathbf{P}_N$, thus $c' + c_{\phi_1}$ is very ample.

c. We write $c_\phi = c + (m+1)c_{\phi_1}$, $c_\psi = (m+1)c_{\phi_1}$, where ϕ and ψ are immersions. Then $c = c_\phi - c_\psi$.

Existence.

a. We check that if $c_{\phi_1} = c_{\phi_2}$, then $h_{\phi_1} = h_{\phi_2} + O(1)$. We consider for this the complete linear system, i.e. the vector space $\Gamma = H^0(V, E)$ of all the global sections of the line bundle E associated to c_{ϕ_1}. The ϕ_i correspond

to bases of sub-vector spaces of Γ. We have already seen that a change of basis or of subspace does not change h_ϕ.
b. We write $c \in \mathrm{Pic}(V)$ in the form $c = c_\phi - c_\psi$, and we would be able to put $h_c = h_\phi - h_\psi$ if we could verify that $c = c_{\phi'} - c_{\psi'} \Rightarrow h_{\phi'} - h_{\psi'} = h_\phi - h_\psi + O(1)$. But as $c_\phi + c_{\psi'} = c_{\phi'} + c_\psi$ one has $h_{\phi\otimes\psi'} = h_{\phi'\otimes\psi} + O(1)$, from a), thus $h_\phi + h_{\psi'} = h_{\phi'} + h_\psi + O(1)$.

Exercise. Give another proof of the existence using the following lemma of Mumford: $\mathrm{Pic}(V)$ is presented by its generators c_ϕ, for morphisms $\phi : X \to \mathbf{P}_n$, and the relations:

1) If $i : \mathbf{P}_n \to \mathbf{P}_{n+1}$ is a linear embedding, then $c_\phi = c_{i\phi}$.
2) $c_{\phi\otimes\psi} = c_\phi + c_\psi$.

Functoriality. *Let $f : V \to V'$, let $c' \in \mathrm{Pic}(V')$, and let $c = f^*c' \in \mathrm{Pic}(V)$. Then*

$$h_c = h_{c'} \circ f + O(1).$$

This is immediate by writing, for example, $c' = c_{\phi'} - c_{\psi'}$.

This property can be applied to a product variety $V_1 \times V_2$ where V_1 and V_2 are non-empty. By considering the projections $V_1 \times V_2 \to V_i$, there is a natural inclusion

$$\mathrm{Pic}(V_1) \oplus \mathrm{Pic}(V_2) \subset \mathrm{Pic}(V_1 \times V_2).$$

(In general this is not an equality, there is an extra factor, as we shall see later (§3.2).) Let $c_1 \in \mathrm{Pic}(V_1)$, $c_2 \in \mathrm{Pic}(V_2)$, and $c = c_1 + c_2 \in \mathrm{Pic}(V_1 \times V_2)$. For $x_i \in V_i(K), (i = 1, 2)$, we have

$$h_c(x_1, x_2) = h_{c_1}(x_1) + h_{c_2}(x_2) + O(1).$$

Examples.
1). To any morphism $f : \mathbf{P}_n \times \mathbf{P}_m \to \mathbf{P}_N$ we associate a pair of integers (d_1, d_2) (corresponding to the degrees of f expressed in terms of multihomogeneous polynomials). Here there is no extra factor: $\mathrm{Pic}(\mathbf{P}_N) = \mathbf{Z}$, $\mathrm{Pic}(\mathbf{P}_n \times \mathbf{P}_m) = \mathbf{Z} \times \mathbf{Z}$, and we get

$$h(f(x, y)) = d_1 h(x) + d_2 h(y) + O(1).$$

2). Let E be an elliptic curve with origin 0, m an integer ≥ 1, $c_1 = m(0)$. (For $m \geq 3$, c_1 is a very ample divisor. For $m = 3$, it gives the standard plane cubic immersion; for $m = 4$, it gives an intersection of two quadrics in $\mathbf{P}_3$.)

We cannot apply the preceding formula to the composition law $s : E \times E \to E$, for s^*c_1 does not belong to $\mathrm{Pic}(E) \oplus \mathrm{Pic}(E)$. But one can take the morphism $E \times E \to E \times E$, $(x,y) \mapsto (x+y, x-y)$. The inverse image of $c_1 \otimes 1 + 1 \otimes c_1$ is $2(c_1 \otimes 1 + 1 \otimes c_1)$, as one can see, either geometrically or by analysis, from formulae for $\wp(u+v).\wp(u-v)$, or for the 4 Jacobi theta functions : $\theta_i(u+v).\theta_i(u-v)$ is an homogeneous polynomial of degree $(2,2)$ in the $\theta_j(u), \theta_j(v)$, $i, j \in \{1,2,3,4\}$ (see e.g. [Wh] p.456, exerc. 8, and p. 488, exerc. 3). For the associated height on $E \times E$ we deduce formulae of the type (cf. §§3.3, 3.6)

$$h(x+y, x-y) = 2h(x) + 2h(y) + O(1).$$

3). Let d be a positive integer, $[d] : E \to E$ the morphism $x \mapsto dx$, and c a multiple of (0). Then, $[d]^*c = d^2 c$. Whence

$$h(dx) = d^2 h(x) + O(1).$$

2.9. $h_c = O(1) \Leftrightarrow c$ is of finite order (number fields).

The function h_c has real values (positive, negative, or zero). If c is of finite order in $\mathrm{Pic}(V)$, we have $h_c = O(1)$. Conversely:

Theorem. *If V is non-singular and projective and K is a number field, then $h_c = O(1)$ implies that c is of finite order in* $\mathrm{Pic}(V)$.

The proof will be given in §3.11, as an application of Néron's quadraticity theorem.

2.10. Positivity of the height.

We shall show that if c is ample, then h_c is bounded below (that is, h_c is positive up to $O(1)$). In the non-singular case, if $c \in \mathrm{Pic}(V)$ corresponds to a positive divisor Δ, then $h_c \geq 0$ (up to $O(1)$) on $V - \mathrm{Supp}(\Delta)$.

The general statement is the following:

Theorem. *Let $c \in \mathrm{Pic}(V)$, E a line bundle associated to c, and s a global section of E. Then, $h_c \geq 0$ (up to $O(1)$) outside the variety of zeros of s.*

Proof.

There are two immersions ϕ' and ϕ'' such that $c = c_{\phi''} - c_{\phi'}$, thus there are two very ample line bundles E' and E'' such that $E \otimes E' \cong E''$. We must show that $h_{c''} \geq h_{c'} + O(1)$ outside the zeros of the section s $(c' = c_{\phi'}, c'' = c_{\phi''})$.

Let $s_0, \ldots, s_n$ be a basis of the global sections of E' ; we have

$$h_{c'}(x) = h(s_0(x), \ldots, s_n(x)).$$

Complete the set $ss_0, \ldots, ss_n$ of global sections of E'' to a basis $(ss_0, \ldots, ss_n, \sigma_1, \ldots,)$ of $\Gamma(V, E'')$. Then

$$h_{c''}(x) = h(ss_0(x), \ldots, ss_n(x), \sigma_1(x), \ldots).$$

At a point where $s(x) \neq 0$, we have

$$h_{c''}(x) \geq h(ss_0(x), \ldots, ss_n(x)),$$

and by homogeneity

$$h_{c'}(x) = h(ss_0(x), \ldots, ss_n(x)),$$

whence the result.

If c corresponds to a positive divisor Δ, then h_c *is positive modulo* $O(1)$ *outside the base points and base components of the linear system* $|\Delta|$, that is to say outside of the intersection of the $\Delta' \geq 0$, $\Delta' \sim \Delta$. In particular, if c is generated by its global sections then $h_c \geq 0$ everywhere (this was clear *a priori* : $h_c = h_\phi + O(1)$).

2.11. Divisors algebraically equivalent to zero.

Let V be a non-singular projective variety, C an irreducible curve, and Δ a divisor on $C \times V$ without a vertical component. For $\lambda \in C$ let

$$V_\lambda = \Delta \cap (\{\lambda\} \times V).$$

In terms of divisor classes, V_λ is the class of the inverse image of Δ by the map $V \to C \times V, x \mapsto (\lambda, x)$. We then obtain a 1-parameter family of divisors. A divisor is said to be *algebraically equivalent to zero* if its class in $\mathrm{Pic}(V)$ belongs to the subgroup $\mathrm{Pic}^\circ(V)$ generated by the divisors $V_\lambda - V_\mu, \lambda, \mu \in C$ (for all possible choices of C, Δ, λ, μ).

The simplest case is that where V is a non-singular projective curve: $\mathrm{Pic}^\circ(V)$ is just the group of divisors of degree 0. The ample divisors are those of degree > 0.

Theorem. *Let c and c_1 be two elements of* Pic(V), *with* $c \in \mathrm{Pic}^\circ(V)$ *and* c_1 *ample. Then for all* $\epsilon > 0$, *we have*

$$|h_c| \leq \epsilon h_{c_1} + O(1).$$

If c_1 and c_2 are ample, then $mc_2 - c_1$ is ample for sufficiently large m, and we deduce that $h_{c_1} \asymp h_{c_2} + O(1)$. Thus if the formula is true for one choice of c_1 it is true for all c_1. We use the following result.

Theorem. *If* $c \in \mathrm{Pic}^\circ(V)$ *and if* c_1 *is ample, then* $c + c_1$ *is ample.*

In other words, ampleness descends to the Néron-Severi group of V:

$$\mathrm{NS}(V) = \mathrm{Pic}(V)/\mathrm{Pic}^\circ(V).$$

This can be proved using the Nakai-Moishezon criterion (see [H, Ch. V, Th. 1.10], for the case of surfaces, and [K], for the general case): a positive divisor D is ample if and only if for all irreducible sub-varieties Δ one has $D^\delta.\Delta > 0$, where δ is the dimension of Δ (in the case of a surface, these conditions become $D^2 > 0$, and $D.C > 0$ for all irreducible curves C on the surface).

We now prove the theorem on $|h_c|$. As $c \in \mathrm{Pic}^\circ(V)$, we have $nc \in \mathrm{Pic}^\circ(V)$ for all $n \in \mathbf{Z}$; therefore $nc + c_1$ is ample and $nh_c + h_{c_1}$ is positive modulo $O(1)$. One bounds h_c from above by choosing $n < -1/\epsilon$, and one bounds $-h_c$ from above by choosing $n > 1/\epsilon$.

We have thus shown $h_c(x) = o(h_{c_1}(x))$, as $h_{c_1}(x) \to \infty$. Néron's theory gives a more precise statement:

Theorem (Néron). $|h_c(x)| \leq O(h_{c_1}^{1/2} + 1)$.

The proof will be given in §3.10.

2.12. Example-exercise: projective plane blown up at a point.

We consider the surface V obtained by blowing up the point O=(0,0,1) of the projective plane $\mathbf{P}_2$. That is to say, $V \subset \mathbf{P}_2 \times \mathbf{P}_1$ is the closure of the graph of

$$\mathbf{P}_2 - \{(0,0,1)\} \to \mathbf{P}_1$$

$$(x, y, z) \mapsto (x, y).$$

Taking homogeneous coordinates (x, y, z) on $\mathbf{P}_2$ and (u, v) on $\mathbf{P}_1$, one may define V in $\mathbf{P}_2 \times \mathbf{P}_1$ by the equation

$$xv - yu = 0.$$

Let $\pi : V \to \mathbf{P}_2$ be the surjection induced from the projection $\mathbf{P}_2 \times \mathbf{P}_1 \to \mathbf{P}_2$. Then π is an isomorphism except at the inverse image of the point O: we have replaced O by the slopes x/y of the lines through O.

The group $\mathrm{Pic}(V)$ is $\mathbf{Z} \times \mathbf{Z}$ with basis L, D defined as follows:

L is the blow up of O, i.e. $\pi^{-1}(\mathrm{O})$.

D is the class of the inverse image in V of a line of $\mathbf{P}_2$ not going through O.

If $\tilde{\Delta}$ is a line of $\mathbf{P}_2$ through O, its inverse image by π is $L + \Delta$, where Δ is the strict transform of $\tilde{\Delta}$ ([H], Ch5, §3). The class of Δ is thus equal to $D - L$.

One checks (e.g. by the Nakai-Moishezon criterion [H, Ch.5, Theorem 1.10]) that the *ample* divisor classes are the linear combinations,

$$c = aD + b(D - L) \sim aD + b\Delta, \text{ with } a, b > 0.$$

Intersection numbers are readily computed from the formulae ([H], Ch.5, Prop. 3.2):

$$D^2 = 1, D.L = 0, L^2 = -1.$$

Let $c = aD + b(D - L)$ be an ample divisor, and let h_c be the corresponding logarithmic height on $V(K)$. Let $H_c = \exp h_c$. Since c is ample, we know that the number of rational points $x \in V(\mathbf{Q})$ with $H_c(x) \leq X$ is finite for every X. The point of the exercise is to *determine the growth* of this number as $X \to \infty$. The result is:

The number $N(X)$ of $x \in V(\mathbf{Q})$ such that $H_c(x) \leq X$ is $\asymp X^\gamma$, where $\gamma = \max(2/a, 2/b)$.

(The symbol $\asymp$ means that, for large enough X, the ratios of the two numbers lies between c_1 and c_2 with c_i positive. A more precise result cannot be expected since h_c is defined only up to the addition of a bounded function.)

(a) Show that the number of $x \in L(\mathbf{Q})$ with $H_c(x) \leq X$ is $\asymp X^{2/b}$.
[Note that $c.L = b$, then apply Schanuel's theorem to L.]

(b) Show that the number of $x \in \Delta(\mathbf{Q})$ with $H_c(x) \leq X$ is $\asymp X^{2/a}$. Deduce that $N(X) \gg X^\gamma$.

(c) Show that $N(X) \ll$ the number of $(x, y, z) \in \mathbf{P}_2(\mathbf{Q})$, $(x, y, z) \neq (0, 0, 1)$ with

$$H(x, y, z)^a H(x, y)^b \leq X.$$

We normalise, as usual, a point $(x, y, z) \in \mathbf{P}_2(\mathbf{Q})$ so that $x, y, z \in \mathbf{Z}$ and $\gcd(x, y, z) = 1$. We may then select the heights in (c) to be

$$H(x, y, z) = \sup(|x|, |y|, |z|),$$

$$H(x, y) = \sup(\frac{|x|}{d}, \frac{|y|}{d}), \text{ where } d = \gcd(x, y).$$

(d) Let $C \subset V$ be the strict transform by π of the reducible curve $xyz = 0$ in $\mathbf{P}_2$. Show that there are $\ll X^\gamma$ points $P \in C(\mathbf{Q})$ with $H_c(P) \leq X$.

[Use Schanuel's theorem.]

(e) Show that if $P \in (V - L - C)(\mathbf{Q})$ with $H_c(P) < X$ then $H(x, y)$ ($= Y$, say) $\leq X^{1/(a+b)}$, $|z| \leq X^{1/a}Y^{-b/a}$ and $d \leq X^{1/a}Y^{-1-b/a}$.

Deduce that the number of $P \in (V - L - C)(\mathbf{Q})$ with $H_c(P) \leq X$ is

$$\ll X^{2/a} \sum_{Y \leq X^{1/(a+b)}} Y^{-2b/a}.$$

(f) Show that the sum in (e) has the upper bound

$$X^{2/a} \sum_{Y \leq X^{1/(a+b)}} Y^{-2b/a} \ll \begin{Bmatrix} X^{2/a} + X^{3/(a+b)}, & a \neq 2b \\ X^{2/a} \log X, & a = 2b \end{Bmatrix} \ll X^\gamma,$$

and deduce the result stated above.

The interest of this exercise is to give an example where one can evaluate the rate of growth of the number of points on V with $H_c \leq X$ and see how it varies with c. It might be interesting to do the same for other varieties, for instance cubic surfaces in $\mathbf{P}_3$ (cf. [MFT]).

Notice that for all c, we found essentially *powers of* X. For abelian varieties Néron's theory (cf. Chapter 3 and §4.5) gives instead *powers of* $\log X$. Problem: can one find a growth which is intermediate between powers of X and powers of $\log X$? Nothing is known on such questions.

3. NORMALISED HEIGHTS

3.1. Néron-Tate normalisation.

Lemma. (Tate). *Let S be a set and $\pi : S \to S$ a map. Let f be a real-valued function on S such that $f \circ \pi = \lambda f + O(1)$, with $\lambda > 1$. Then there is a unique function $\tilde{f}$ on S such that*
(a) $\tilde{f} = f + O(1)$
(b) $\tilde{f} \circ \pi = \lambda \tilde{f}$
and we have

$$\tilde{f}(x) = \lim_{n\to\infty} (1/\lambda^n) f(\pi^n x), \textit{for every } x \in S.$$

We shall show (a) in the following more precise form: if c is such that $|f(\pi x) - \lambda f(x)| \leq c$, then

$$|\tilde{f}(x) - f(x)| \leq c/(\lambda - 1).$$

Proof.
Let $c > 0$ be such that $|f(\pi x) - \lambda f(x)| \leq c$. We have

$$|(1/\lambda^n) f(\pi^n x) - (1/\lambda^{n-1}) f(\pi^{n-1} x)| \leq c/\lambda^n,$$

so the condition $\lambda > 1$ shows that the sequence $(1/\lambda^n) f(\pi^n x)$ is convergent. Let $\tilde{f}(x)$ be its limit. Property (b) is clear, and

$$|\tilde{f}(x) - f(x)| \leq \sum_{n=1}^{\infty} c/\lambda^n = c/(\lambda - 1).$$

Observe that if f is bounded then $\tilde{f} = 0$. Thus if $f = g + O(1)$, then $\tilde{f} = \tilde{g}$.

Uniqueness. If $g = f + O(1)$ with $g \circ \pi = \lambda g$, we have from the above that $\tilde{g} = \tilde{f}$, and $\tilde{g} = g$, thus $g = \tilde{f}$.

Exercise. (Alternative proof) Let E be the Banach space of bounded real-valued functions on S equipped with the sup norm. Show that the operator $A : E \to E$, $f \mapsto f \circ \pi$ is such that $||A|| \leq 1$ and hence that $1 - (1/\lambda)A$ is invertible. If ϕ satisfies

$$(1 - (1/\lambda)A)\phi = -f + (1/\lambda) f \circ \pi,$$

show that $\tilde{f} = f + \phi$ has the properties (a) and (b) of Tate's lemma.

Functoriality of $\tilde{f}$.

Let $\sigma : S \to S'$ be a map, and let $\pi : S \to S$, $\pi' : S' \to S'$ be such that $\pi' \circ \sigma = \sigma \circ \pi$. Let $f' : S' \to \mathbf{R}$ have $f' \circ \pi' = \lambda f' + O(1)$, and let $f = f' \circ \sigma$.

$$\begin{array}{ccccc} S & & \xrightarrow{\sigma} & & S' \\ \downarrow{\scriptstyle \pi} & & & & \downarrow{\scriptstyle \pi'} \\ S & & \xrightarrow{\sigma} & & S' \\ & \searrow{\scriptstyle f} & & \swarrow{\scriptstyle f'} & \\ & & \mathbf{R} & & \end{array}$$

Then $f \circ \pi = \lambda f + O(1)$, and $\tilde{f} = \tilde{f}' \circ \sigma$.

Proof: by the uniqueness.

Commutativity property.

Let S, π, f, λ be as in Tate's lemma (in particular, $\lambda > 1$). Let π_α be endomorphisms of S commuting with π, and λ_α real numbers, such that

$$f \circ \pi_\alpha = \lambda_\alpha f + O_\alpha(1).$$

Then

$$\tilde{f} \circ \pi_\alpha = \lambda_\alpha \tilde{f}.$$

In particular, if π_α are commuting endomorphisms of a set S, and f is a real-valued function on S such that $f \circ \pi_\alpha = \lambda_\alpha f + O_\alpha(1)$ with $\lambda_\alpha \in \mathbf{R}$, where one of the λ_α's is > 1, then there is a unique $\tilde{f}$ such that $\tilde{f} \circ \pi_\alpha = \lambda_\alpha \tilde{f}$ for all α.

Proof: this follows from the functoriality applied to

$$\begin{array}{ccc} S & \xrightarrow{\pi_\alpha} & S \\ \downarrow{\scriptstyle \pi} & & \downarrow{\scriptstyle \pi} \\ S & \xrightarrow{\pi_\alpha} & S. \end{array}$$

Normalised heights (Néron [N4], Tate).

Let K be a field equipped with a product formula, X a projective variety over $\bar{K}$, $S = X(\bar{K})$, π a morphism $X \to X$, and $c \in \mathrm{Pic}(X)$.

Suppose that $\pi^* c = \lambda c$ with $\lambda > 1$. By §2.8, we have $h_c(\pi x) = \lambda h_c(x) + O(1)$ on $X(\bar{K})$. Hence by Tate's lemma, there is a unique function

$\tilde{h}_c$ such that $\tilde{h}_c = h_c + O(1)$ and $\tilde{h}_c(\pi x) = \lambda \tilde{h}_c(x)$. This is the normalised logarithmic height.

Example. Projective space $\mathbf{P}_n$.

Let $d \in \mathbf{Z}$ and $\pi_d(x_0, ..., x_n) = (x_0^d, ..., x_n^d)$. Let c be the natural generator of $\mathrm{Pic}(\mathbf{P}_n) = \mathbf{Z}$ (corresponding to the sheaf $\mathbf{O}(1)$ and to the usual height on projective space). Then $\pi_d^* c = dc$. We recover the standard height since

$$h(x_0^d, ..., x_n^d) = dh(x_0, ..., x_n).$$

3.2. Abelian varieties.

For any abelian variety A, one has an exact sequence

$$0 \to \mathrm{Pic}^\circ(A) \to \mathrm{Pic}(A) \to \mathrm{NS}(A) \to 0,$$

where $\mathrm{Pic}^\circ(A)$ is the Picard variety of A (divisor classes algebraically equivalent to zero) and $\mathrm{NS}(A)$ (Néron -Severi group) is a finitely generated group.

[Over $\mathbf{C}$, we may see this by transcendental methods: we take the cohomology of the exact sequence of sheaves

$$0 \longrightarrow \mathbf{Z} \xrightarrow{2\pi i} \mathbf{O}_{\mathrm{hol}} \xrightarrow{\exp} \mathbf{O}^*_{\mathrm{hol}} \longrightarrow 0$$

The H^0 terms provide the exact sequence of groups $0 \to \mathbf{Z} \to \mathbf{C} \to \mathbf{C}^* \to 0$, because global holomorphic functions on connected projective varieties are constant. We have then the exact sequence

$$0 \to \mathrm{H}^1(A, \mathbf{Z}) \to \mathrm{H}^1(A, \mathbf{O}_{\mathrm{hol}}) \to \mathrm{H}^1(A, \mathbf{O}^*_{\mathrm{hol}}) \to \mathrm{H}^2(A, \mathbf{Z}) \to \dots$$

This gives an injection of $\mathrm{H}^1(A, \mathbf{O}_{\mathrm{hol}})/\mathrm{H}^1(A, \mathbf{Z})$ in $\mathrm{H}^1(A, \mathbf{O}^*_{\mathrm{hol}})$ where the cokernel (which may be identified with $\mathrm{NS}(A)$) is a subgroup of $\mathrm{H}^2(A, \mathbf{Z})$. Further

$$\mathrm{H}^1(A, \mathbf{O}^*_{\mathrm{hol}}) \cong \mathrm{Pic}(A),$$

$\mathrm{H}^1(A, \mathbf{O}_{\mathrm{hol}})$ is a $\mathbf{C}$-vector space of dimension $\dim A$, and $\mathrm{H}^1(A, \mathbf{Z})$ is a lattice of rank $2 \dim A$. We thus have an analytic description of the Picard variety as a quotient

$$\mathrm{Pic}^\circ(A) \cong \mathrm{H}^1(A, \mathbf{O}_{\mathrm{hol}})/\mathrm{H}^1(A, \mathbf{Z}).]$$

For $d \in \mathbf{Z}$, we write $\pi_d = [d] = d$ for the endomorphism $x \mapsto dx$. In particular, $[-1]$ is the natural symmetry of the variety, inducing an involution on $\mathrm{Pic}(A) : c \mapsto c' = [-1]^* c$. If c is represented by a divisor D, c' is represented by the image of D under $x \mapsto -x$. We say that c is symmetric if $c = c'$, antisymmetric if $c = -c'$. The symmetry $[-1]$ acts trivially on the Néron-Severi group $\mathrm{NS}(A)$ and induces the natural symmetry $c \mapsto -c$ on $\mathrm{Pic}^\circ(A)$.

We need the 'theorem of the cube':

Theorem. *Let X, Y, Z be three projective varieties; any element $c \in \mathrm{Pic}(X \times Y \times Z)$ which induces 0 on $X \times Y$, $X \times Z$, $X \times Z$ is zero.*

For the proof, see [Mu], §6 and §10.

Consider now the following morphisms from $A \times A \times A$ to A:

$$s_{123}(x, y, z) = x + y + z,$$

$$s_{12}(x, y, z) = x + y, \quad s_{23}(x, y, z) = y + z, \quad s_{13}(x, y, z) = x + z,$$

$$s_1(x, y, z) = x, \quad s_2(x, y, z) = y, \quad s_3(x, y, z) = z.$$

Theorem. *Let $c \in \mathrm{Pic}(A)$. In $\mathrm{Pic}(A \times A \times A)$ we have*

$$s_{123}^* c - s_{12}^* c - s_{13}^* c - s_{23}^* c + s_1^* c + s_2^* c + s_3^* c = 0.$$

Proof.

Let

$$\sigma = s_{123}^* c - s_{12}^* c - s_{13}^* c - s_{23}^* c + s_1^* c + s_2^* c + s_3^* c.$$

By symmetry and the theorem of the cube, it suffices to show that σ is zero on $0 \times A \times A$. As $s_{123}(0, y, z) = y + z = s_{23}(0, y, z)$, on $0 \times A \times A$ we have $s_{123} = s_{23}$, and also $s_{12} = s_2$, $s_{13} = s_3$, $s_1 = 0$. Whence the theorem.

Functions of degree ≤ 2 between abelian groups.

If G_1 and G_2 are two abelian groups and $F : G_1 \to G_2$ is a map such that $F(0) = 0$, one says that

a) *F is of degree* ≤ 1 if $F(x+y) - F(x) - F(y) = 0$.

b) *F is of degree* ≤ 2 if $F_1(x, y) = F(x+y) - F(x) - F(y)$ is additive in x and y, or equivalently, if the second difference $F_2(x, y, z) = F(x + y + z) - F(x + y) - F(x + z) - F(y + z) + F(x) + F(y) + F(z)$ is zero.

Theorem. *Let X be an algebraic variety, and A an abelian variety. Let $c \in \mathrm{Pic}(A)$. Then the map of abelian groups,*

$$\mathrm{Hom}(X, A) \to \mathrm{Pic}(X), \quad f \mapsto f^* c,$$

has degree ≤ 2.

Proof.

We deduce this from the preceding theorem by a simple functoriality argument. Let, $f_i : X \to A$, $i = 1, 2, 3$, be three morphisms, and

$$f = (f_1, f_2, f_3) : X \to A \times A \times A.$$

Take as before

$$\sigma = s_{123}^* c - s_{12}^* c - s_{13}^* c - s_{23}^* c + s_1^* c + s_2^* c + s_3^* c.$$

We have

$$f^*\sigma = (f_1+f_2+f_3)^*c-(f_1+f_2)^*c-(f_1+f_3)^*c-(f_2+f_3)^*c+f_1^*c+f_2^*c+f_3^*c.$$

Since $\sigma = 0$, we have $f^*\sigma = 0$; hence the theorem.

This implies the following formula for $[d]^*c$, where $c' = [-1]^*c$:

Corollary (Weil , [Mu2]). *If $d \in \mathbf{Z}$ and $c \in \mathrm{Pic}(A)$, we have*

$$[d]^*c = \frac{d(d+1)}{2}c + \frac{d(d-1)}{2}c'.$$

In particular:

$$[d]^*c = \begin{cases} d^2c, & \text{if } c \text{ is symmetric;} \\ dc, & \text{if } c \text{ is antisymmetric.} \end{cases}$$

Proof.

The map $\mathbf{Z} \to \mathrm{Pic}(A)$, $d \mapsto [d]^*c$, has degree ≤ 2 by the previous theorem, therefore the result follows from the next lemma.

Lemma. *Let G be an abelian group, and $F : \mathbf{Z} \to G$ a function of degree ≤ 2 with $F(0) = 0$. Then*

$$F(n) = \frac{n(n+1)}{2}F(1) + \frac{n(n-1)}{2}F(-1).$$

Proof of lemma.

Taking the difference between the two sides, we reduce to the case where $F(1) = F(-1) = F(0) = 0$. As

$$F_1(-1, 1) = F(0) - F(-1) - F(1) = 0,$$

and F_1 is biadditive, we have $F_1 = 0$, therefore F is additive. But $F(1) = 0$, therefore $F = 0$.

Exercises: quadratic functors on projective varieties.

Let $X \to F(X)$ be a contravariant functor from the category C of pointed (equipped with a base point) smooth connected projective varieties over $\bar{K}$ to the category of abelian groups. We assume that if P is a point, then $F(P) = 0$.

1) Let X, Y, Z be three varieties in C. Let us identify X, Y, Z with subvarieties of the products $X \times Y, X \times Y \times Z$, by means of the base points. Show that we have the decompositions

$$F(X \times Y) = F(X) \oplus F(Y) \oplus F_2(X,Y),$$

$$F(X \times Y \times Z) =$$

$$F(X) \oplus F(Y) \oplus F(Z) \oplus F_2(X,Y) \oplus F_2(X,Z) \oplus F_2(Y,Z) \oplus F_3(X,Y,Z),$$

where $F_2(X,Y), F_3(X,Y,X)$ are functors in X, Y and X, Y, Z, respectively. (We say that F *has degree* ≤ 1, or is *additive* if $F_2 = 0$; F *has degree* ≤ 2 if $F_3 = 0$, etc.)

2) Show that the contravariant functors $H^1(X)$ (the usual complex cohomology over $\mathbf{Z}$), $H^1(X, \mathbf{O}_X)$, $\mathrm{Pic}^\circ(X)$ are additive.

3) Show that $H^2(X, \mathbf{Z})$ and the coherent cohomology $H^2(X, \mathbf{O}_X)$ are functors of degree ≤ 2. (Over $\mathbf{C}$, the extra factor can be described by the Künneth formula:

$$F_2(X,Y) = H^1(X) \otimes H^1(Y).)$$

More generally, show that $H^q(X, \mathbf{Z})$ is a functor of degree $\leq q$.

4) Show that the Picard functor $\mathrm{Pic}(X)$ has degree ≤ 2 by

(a) using the theorem of the cube,

or by

(b) over $\mathbf{C}$, using the complex analytic exact sequence

$$H^1(X, \mathbf{O}_X) \to \mathrm{Pic}(X) \to H^2(X, \mathbf{Z}),$$

and that $H^1(X, \mathbf{O}_X)$, $H^2(X, \mathbf{Z})$ have degree ≤ 2.

We may then normalise the heights h_c, $c \in \mathrm{Pic}(A)$:

Theorem. (Néron-Tate). *Let A be an abelian variety over $\bar{K}$. There is a unique function $c \mapsto \tilde{h}_c$ on $\mathrm{Pic}(A)$ with values in the space of real valued functions on $A(\bar{K})$ such that,*

1) $\tilde{h}_c(x) = h_c(x) + O(1)$.

2) Additivity: $\tilde{h}_{c_1+c_2} = \tilde{h}_{c_1} + \tilde{h}_{c_2}$.

3) Functoriality: for all endomorphisms $\phi : A \to A$, we have

$$\tilde{h}_{\phi^* c} = \tilde{h}_c \circ \phi \quad \textit{for} \quad c \in \mathrm{Pic}(A).$$

Further if B is another abelian variety and $\psi : B \to A$ is a homomorphism, then

$$\tilde{h}_{\psi^* c} = \tilde{h}_c \circ \psi \quad \textit{for all } c \in \mathrm{Pic}(A).$$

Proof.

1) Suppose that c is symmetric. Consider the homothety $\phi = [2]$. From the corollary above, we know that $\phi^* c = 4c$. There is then a unique $\tilde{h}_c$ satisfying 1) and $\tilde{h}_c(2x) = 4\tilde{h}_c(x)$. The uniqueness of $\tilde{h}_c$ follows from this and 2),3).

If ϕ is any endomorphism, then $\phi^* c$ is symmetric, and the commutativity lemma (§3.1) applied to the diagram,

$$\begin{array}{ccc} A & \xrightarrow{2} & A \\ \downarrow{\scriptstyle\phi} & & \downarrow{\scriptstyle\phi} \\ A & \xrightarrow{2} & A \end{array}$$

gives $\tilde{h}_{c_\phi} = \tilde{h}_c \circ \phi$.

The additivity is clear, by the uniqueness.

2) Suppose that c is antisymmetric. Then $\phi^* c = 2c$ and one uses the same argument.

3) In the general case, we have $2c = (c + c') + (c - c')$ and we put $\tilde{h}_c = (\tilde{h}_{c+c'} + \tilde{h}_{c-c'})/2$.

3.3. Quadraticity of $\tilde{h}_c$ on abelian varieties.

Theorem. *Let $c \in \mathrm{Pic}(A)$. The normalised height $\tilde{h}_c$ is a function of degree ≤ 2 on $A(\bar{K})$.*

Proof.

Let

$$\Delta_2 \tilde{h}_c(x, y, z) = \tilde{h}_c(x + y + z) - \tilde{h}_c(x + y) - \tilde{h}_c(x + z) - \tilde{h}_c(y + z) + \\ + \tilde{h}_c(x) + \tilde{h}_c(y) + \tilde{h}_c(z).$$

We must show that $\Delta_2 \tilde{h}_c = 0$. Let $s_{123}, s_{12}, s_1, \ldots : A \times A \times A \to A$ be the seven morphisms given in §3.2. By the functorial property of normalised heights, we have

$$\tilde{h}_{s_{123}^* c}(x, y, z) = \tilde{h}_c \circ s_{123}(x, y, z) = \tilde{h}_c(x + y + z),$$

and similarly for the other terms in $\Delta_2 \tilde{h}_c$. As $\sigma = s_{123}^* c - \sum s_{12}^* c + \sum s_1^* c = 0$ (§3.2), we deduce $\tilde{h}_\sigma = 0$. By additivity, $\tilde{h}_\sigma = \tilde{h}_{s_{123}^* c} - \sum \tilde{h}_{s_{12}^* c} + \sum \tilde{h}_{s_1^* c}$, whence the theorem.

Let $B_c(x, y)$ be the bilinear form associated to $\tilde{h}_c$:

$$B_c(x, y) = \tilde{h}_c(x + y) - \tilde{h}_c(x) - \tilde{h}_c(y).$$

If c is antisymmetric, the function $\tilde{h}_c(x)$ is of degree ≤ 2 and odd: $\tilde{h}_c(-x) = -\tilde{h}_c(x)$; so $B_c(x, y)$ is biadditive and antisymmetric:

$$B_c(-x, -y) = -B_c(x, y),$$

therefore $B_c = 0$ and $\tilde{h}_c$ is additive.

If c is symmetric, $\tilde{h}_c$ is a quadratic function:

$$\tilde{h}_c(x) = \frac{1}{2} B_c(x, x).$$

Hence:

Theorem. *Let A be an abelian variety, $c \in \mathrm{Pic}(A)$. Then:*
1). If c is symmetric, the normalised height $\tilde{h}_c$ is quadratic on $A(\bar{K})$.
2). If c is antisymmetric, $\tilde{h}_c$ is $\mathbf{Z}$-linear (additive).

In general, the difference, $f(x) = \tilde{h}_c(x) - \frac{1}{2} B_c(x, x)$ is additive. It remains to explain how this additive part can be computed from the quadratic part. For this we recall some properties of the dual abelian variety.

3.4. Duality and Poincaré divisors.

Let A be an abelian variety. The dual abelian variety A' has the following properties (which characterise it, up to isomorphism in characteristic zero, and up to purely inseparable isogeny in characteristic p).

The group A' may be identified with $\mathrm{Pic}^\circ(A)$. This identification is given in the following way. On $A \times A'$ there is a canonical element $P \in \mathrm{Pic}(A \times A')$ (Poincaré divisor class) with the following properties:

a) P induces 0 on $A \times 0$ and on $0 \times A'$.
b) For $a' \in A'$, the divisor class P induces on $A \times \{a'\}$ (identified with A) the element a', viewed as an element of $\mathrm{Pic}^\circ(A)$.

The equivalence class of such a P is unique. Exchanging the roles of A and A' leaves P invariant, i.e. identifies A'' with A.

Example. The dual of the elliptic curve E is E itself. The Poincaré divisor is the divisor $\Delta - \{0\} \times E - E \times \{0\}$ on $E \times E$, where Δ is the diagonal. We obtain the usual identification of E with $\mathrm{Pic}^\circ(E)$: $x \mapsto$ class of $(x) - (0)$.

The uniqueness of the Poincaré divisor class shows that it is symmetric

$$P = P' = [-1]^* P.$$

We associate to $P \in \mathrm{Pic}(A \times A')$ the normalised height $\tilde{h}_P$ on $A(\bar{K}) \times A'(\bar{K})$.

Proposition. *The height $\tilde{h}_P$ is a quadratic form which is trivial on the two factors:*

$$\tilde{h}_P(x, 0) = 0, \text{ for } x \in A(\bar{K}), \quad \tilde{h}_P(0, y) = 0 \text{ for } y \in A'(\bar{K}).$$

This follows from property a) of the Poincaré divisor class.

Let B_P be the bilinear form associated to $\tilde{h}_P$. As $(x, y) = (x, 0)+(0, y)$, we have

$$\begin{aligned} \tilde{h}_P(x, y) &= \tilde{h}_P(x, 0) + \tilde{h}_P(0, y) + B_P((x, 0), (y, 0)) \\ &= B_P((x, 0), (y, 0)), \end{aligned}$$

therefore $\tilde{h}_P(x, y)$ is additive in x, y. We then have a *canonical bilinear form* on $A(\bar{K}) \times A'(\bar{K})$.

We now use property b). To $y \in A'(\bar{K})$ (i.e. $y \in \mathrm{Pic}^\circ(A)$) corresponds the canonical height $\tilde{h}_y$.

Theorem. *For $(x, y) \in A(\bar{K}) \times A'(\bar{K})$, we have*

$$\tilde{h}_y(x) = \tilde{h}_P(x, y).$$

We first prove a general lemma on translations.

Lemma. *Let A be an abelian variety, $c \in \mathrm{Pic}(A)$, $a \in A(\bar{K})$, T_a the translation $x \mapsto x + a$, and $c_a = T_a c = T_{-a}^* c$. Then*

$$\tilde{h}_{c_a}(x) = \tilde{h}_c(x) - B_c(x, a),$$

where B_c is the bilinear form associated to $\tilde{h}_c$.

Proof of the lemma. Define $f(x)$ by

$$f(x) = \tilde{h}_{c_a}(x) - \tilde{h}_c(x) + B_c(x, a).$$

Since

$$\tilde{h}_{c_a}(x) = \tilde{h}_c(x - a) + O(1) = \tilde{h}_c(x) + \tilde{h}_c(-a) + B_c(x, -a) + O(1),$$

we have $f(x) = O(1)$. Hence $f(x)$ is a bounded function of degree ≤ 2 and $f(0) = 0$. Therefore $f = 0$.

Proof of the theorem.

Consider the map $f_y : A \to A \times A'$, $x \mapsto (x, y)$. We have $f_y = T_{0,y} \circ f_0$. From b), $f_y^* P = y \in \mathrm{Pic}^\circ(A)$. Hence $y = f_y^* P = f_0^* T_{0,y}^* P$, and since f_0 is a homomorphism, we have

$$\tilde{h}_y = \tilde{h}_{T_{0,y}^* P} \circ f_0,$$

that is to say

$$\tilde{h}_y(x) = \tilde{h}_{T_{0,y}^* P}(x, 0) = \tilde{h}_P(x, 0) - B(x, y),$$

from the lemma with $a = (0, -y)$. But $\tilde{h}_P(x, 0) = 0$ and $B(x, y) = \tilde{h}_P(x, y)$. Whence the theorem.

Relation between quadratic and linear forms.

Let $c \in \mathrm{Pic}(A)$. We define $\phi_c a = T_a c - c$ for $a \in A$. As $\phi_c a$ is in an algebraic family containing 0, we have $\phi_c a \in \mathrm{Pic}^\circ(A)$. The theorem of the cube shows that $\phi_c : A \to A'$ is a homomorphism. Further, if c is ample, then ϕ_c is an isogeny.

Let $c \in \mathrm{Pic}(A)$. Let B_c be the bilinear form attached to the canonical height $\tilde{h}_c$.

Theorem. *If $x, y \in A(\bar{K})$, we have*

$$B_c(x,y) = -\tilde{h}_P(x, \phi_c y).$$

Proof.
As $\phi_c y = T_y c - c$, we have

$$\tilde{h}_P(x, \phi_c y) = \tilde{h}_{\phi_c y}(x) = \tilde{h}_{T_y c}(x) - \tilde{h}_c(x).$$

But the lemma above gives

$$\tilde{h}_{T_y c}(x) = \tilde{h}_c(x) - B_c(x,y),$$

whence the theorem.

Remark. If one knows $\tilde{h}_P$ on $A(\bar{K}) \times A'(\bar{K})$, one can reconstruct all the canonical heights associated to the elements of $\mathrm{Pic}(A)$. For if c is symmetric, we have from the last theorem

$$\tilde{h}_c(x) = \frac{1}{2} B_c(x,x) = -\frac{1}{2}\tilde{h}_P(x, \phi_c x).$$

In the general case, one writes $2c = (c + c') + (c - c')$, where $c + c'$ is symmetric and $c - c' \in A'(\bar{K})$, since $[-1]$ acts trivially on $\mathrm{NS}(A)$; therefore

$$\begin{aligned} \tilde{h}_c(x) &= \frac{1}{2} B_c(x,x) + \frac{1}{2}\tilde{h}_{c-c'}(x) \\ &= -\frac{1}{2}\tilde{h}_P(x, \phi_c x) + \frac{1}{2}\tilde{h}_P(x, c - c'). \end{aligned}$$

3.5. Example: elliptic curves.

Let E be an elliptic curve, that is an abelian variety with $\dim E = 1$. A divisor $D = \sum n_P . P$ on E is equivalent to zero if and only if $\deg D = \sum n_P = 0$ and $\sum n_P P = 0$ in the group $E = E(\bar{K})$. Hence $\mathrm{Pic}(E) = \mathbf{Z} \times E$, the projection $\mathrm{Pic}(E) \to \mathbf{Z}$ is given by the degree and $\mathrm{Pic}(E) \to E$ by the sum on the elliptic curve; a generator c of $\mathbf{Z}$ is $D = (0)$ (a divisor of degree 1) and the points of E in $\mathrm{Pic}(E)$ are represented by $(a) - (0)$.

The action of the symmetry $[-1] : c \mapsto c'$ is the following: on the factor $\mathbf{Z}$, it is the identity (the degree does not change by any automorphism) and on the factor A it is -1. A divisor class represented by (n, a) is symmetric if and only if $2a = 0$, and it is antisymmetric if and only if $n = 0$. In particular, every element of Pic(E) is a sum of a symmetric and an antisymmetric element.

There are three interesting choices of ample divisors: $c_1 = (0)$, $c_2 = 2(0)$, $c_3 = 3(0)$. By the additivity, we have $\tilde{h}_{c_i} = i\tilde{h}_{c_1}$, ($i = 1, 2, 3$).

The ample divisor c_1 corresponds to a generator of $\mathbf{Z}$ in Pic(E). The linear system corresponding to (0) is just (0).

The linear system corresponding to c_2 has no fixed points (c_2 is generated by its global sections); it consists of pairs of opposite points on the elliptic curve. In the Weierstrass embedding $y^2 = 4x^3 - g_2x - g_3$, c_2 corresponds to $E \to \mathbf{P}_1$ given by $(x, y) \mapsto (1, x)$. Thus $H_2(P) = H(1, x)$.

The divisor c_3 has no fixed points and is very ample. In the Weierstrass embedding, $H_3(P) = H(1, x, y)$. We have

$$H_3(1, x, y) \asymp H_2(1, x)^{3/2}$$

and

$$\begin{aligned} \tilde{h}_{c_1}(P) &= \frac{1}{2} \log H_2(1, x) + O(1) \\ &= \frac{1}{3} \log H_3(1, x, y) + O(1). \end{aligned}$$

Explicit bounds for the $O(1)$ can be given (see [MZ] for $c_4 = 4(0)$; [Z]).

3.6. Exercises on elliptic curves.

Let E be an elliptic curve as above. Let $c = c_1$ be the element of $\mathrm{Pic}(E) \cong \mathbf{Z} \times E$ corresponding to $(1, 0)$. Consider the following morphisms $E \times E \to E$: the projections $p_1(x, y) = x$, $p_2(x, y) = y$, the sum $s(x, y) = x + y$ and the difference $d(x, y) = x - y$. Let

$$\phi = (s, d) : E \times E \to E \times E$$

$$(x, y) \mapsto (x + y, x - y).$$

(1). Show that

$$\phi^*(c \otimes 1 + 1 \otimes c) = 2(c \otimes 1 + 1 \otimes c),$$

that is to say,

$$s^*c + d^*c = 2\{p_1^*c + p_2^*c\}.$$

Deduce that

$$\tilde{h}_c(x + y) + \tilde{h}_c(x - y) = 2\tilde{h}_c(x) + 2\tilde{h}_c(y).$$

[In the Weierstrass embedding $E \to \mathbf{P}_2$, $P \mapsto (1, x(P), y(P))$, the function $F(P,Q) = x(P) - x(Q)$ on $E \times E$ has divisor $\Delta + \Delta_- - 2\{(0) \times E + E \times (0)\}$, where Δ is the diagonal $x = y$ and Δ_- is the divisor $x + y = 0$.]

(We have already seen this formula (§2.8, Example 2) without the tilde's but with an $O(1)$-term; for normalised heights there is no $O(1)$.)

A function on a group which satisfies this property is necessarily quadratic:

(2). Let E be an abelian group, R an abelian group in which 2 is invertible, and $h : E \to R$ a map such that $h(x+y) + h(x-y) = 2h(x) + 2h(y)$. Show that the function $B(x,y) = h(x+y) - h(x) - h(y)$ is $\mathbf{Z}$-bilinear and symmetric, and $h(x) = \frac{1}{2}B(x,x)$.

[The second difference $C(x,y,z)$ (see §3.3) of h equals $B(x+y,z) - B(x,z) - B(y,z)$ which is an odd function of x and is symmetric in x, y, z; but h is even so that $C(-x,-y,-z) = C(x,y,z)$, hence $C = 0$.]

(3). If λ is an endomorphism of E, show that

$$\tilde{h}_c(\lambda x) = (\deg \lambda)\tilde{h}_c(x).$$

(4). Let $a, x \in E(\bar{K})$; identify a with class$((a) - (0))$ in $\mathrm{Pic}^\circ(E)$. Show that $\tilde{h}_a(x) = -B(a,x)$, where B is the bilinear form associated to $\tilde{h}_c$.

3.7. Applications to properties of heights.

Torsion. Let $V = A(\bar{K})/A(\bar{K})_{\mathrm{tors}}$ where $A(\bar{K})_{\mathrm{tors}}$ is the torsion subgroup of $A(\bar{K})$ (as we are over an algebraically closed field, A is divisible). Plainly, V is a $\mathbf{Q}$-vector space and $\tilde{h}_c$ factors through V:

$$\tilde{h}_c(x+t) = \tilde{h}_c(x), \quad \text{for } c \in \mathrm{Pic}(A), x \in A(\bar{K}), t \in A(\bar{K})_{\mathrm{tors}}.$$

[This is a general property of functions of finite degree with values in a $\mathbf{Q}$-vector space.]

Positivity. If $c \in \mathrm{Pic}(A)$ is ample and symmetric, then $\tilde{h}_c(x) \geq 0$ for all $x \in A(\bar{K})$.

Proof.

By §2.10, there is $C \in \mathbf{R}$ such that

$$\tilde{h}_c(x) \geq C, \text{ for all } x \in A(\bar{K}).$$

Thus for all $n > 0$, we have $\tilde{h}_c(nx) \geq C$, therefore $\tilde{h}_c(x) \geq C/n^2$. Letting $n \to \infty$, we get $\tilde{h}_c(x) \geq 0$.

Exercise. Show that the condition "c is symmetric" cannot be dropped.

3.8. Non-degeneracy.

Theorem. *Suppose that A is defined over $\bar{\mathbf{Q}}$. If c is ample, the quadratic part of $\tilde{h}_c$ is a positive non-degenerate form on $V = A(\bar{\mathbf{Q}})/A(\bar{\mathbf{Q}})_{\mathrm{tors}}$.*

We write $\tilde{h}_c = Q_c + L_c$, where Q_c, L_c are, respectively, quadratic and linear on V. We extend these to $V_{\mathbf{R}} = \mathbf{R} \otimes V$.

Let V be a vector space over $\mathbf{Q}$, F a real valued quadratic form on V . We shall say that F is non-degenerate if for all finite dimensional sub-spaces V' of V, F considered as a quadratic form on $V'_{\mathbf{R}}$ is non-degenerate in the usual sense (non-zero discriminant).

If F is positive, this is equivalent to $V_{\mathbf{R}}$ having no non-zero isotropic vectors (but such a condition for V itself is insufficient; for example, in 2 dimensions with coordinates (x, y), the form $(x - \pi y)^2$ is positive, has no non-zero isotropic vector over $\mathbf{Q}$ but acquires some after tensoring with $\mathbf{R}$).

Proof of the theorem.

The quadratic part of $\tilde{h}_c$ is $\frac{1}{2}(\tilde{h}_c + \tilde{h}_{c'})$. As $c + c'$ is symmetric, we may suppose that $c = c'$.

Lemma 1. *If Λ is a finitely generated subgroup of V, then, for any $M \in \mathbf{R}$, there are only finitely many elements λ of Λ such that $\tilde{h}_c(\lambda) \leq M$.*

Proof.

We lift a set of generators of Λ to points $x_1, ..., x_n$ of $A(K)$ with K finite over $\mathbf{Q}$. As c is ample, one of its multiples is very ample, and we may suppose that c itself is so; up to $O(1)$, $\tilde{h}_c$ is the usual height and we then apply Northcott's theorem (§2.4).

Lemma 2. *Let F be a quadratic form in n variables $x_1, ..., x_n$ with real values. Suppose that for all $M \in \mathbf{R}$ there are only finitely many points $(x_1, ..., x_n) \in \mathbf{Z}^n$ with $F(x) \leq M$. Then F is positive and non-degenerate.*

Proof.

The assumption on F implies that F is ≥ 0 on $\mathbf{Z}^n$, hence on $\mathbf{Q}^n$, hence on $\mathbf{R}^n$ by continuity.

We now show that F is non-degenerate. If not, F would come from a form on $\mathbf{R}^{n-1}$ by some projection $p : \mathbf{R}^n \to \mathbf{R}^{n-1}$. Now, $p|_{\mathbf{Z}^n}$ is injective, by the finiteness hypothesis. Therefore $p(\mathbf{Z}^n)$ is not discrete in $\mathbf{R}^{n-1}$, thus there is a sequence of distinct points x^i of $\mathbf{Z}^n$ such that $p(x^i) \to 0$ in $\mathbf{R}^{n-1}$. Then $F(x^i) \to 0$, whence a contradiction with the hypothesis for $M = 1$.

The theorem follows from lemma 1 and lemma 2.

Questions. a) Is there an elliptic curve E over $\mathbf{Q}$ and two non-torsion points x, y of $E(\mathbf{Q})$ such that $B(x, y) = 0$ (where B is the scalar product provided by the Poincaré divisor on $E \times E$)?

b) What is the transcendence degree of the field generated by the values of B? Can it be $< r(r+1)/2$, where $r = \text{rank } E(\mathbf{Q})$?

3.9. Structure of $A(K)$: a preliminary result.

Theorem. *If K is a number field and A an abelian variety defined over K, then $A(K) = F \oplus L$, where F is finite and L is free over $\mathbf{Z}$.*

[We shall see in the next chapter that L is finitely generated (Mordell-Weil theorem).]

Proof.

Let F be the torsion subgroup of $A(K)$, $L = A(K)/F$, and $V_K = \mathbf{Q} \otimes L$.

1) *F is finite.*

We select c to be symmetric and very ample. We have $\tilde{h}_c(x) = 0$ if $x \in F$. Northcott's theorem (§2.4) implies that the number of such x is bounded, thus the order of F is finite.

[We may also prove the finiteness of F by a p-adic argument. Let K_v be the completion of K for v above the prime number p, then $F \subset A(K_v)$. Now, $A(K_v)$ is a compact p-adic Lie group, hence has an open subgroup U which is torsion free. Since F embeds into $A(K_v)/U$, F is finite.]

Each of these two arguments show a little more: if d is a given integer, the elements of $A(\bar{K})_{\text{tors}}$ of degree $\leq d$ over K form a finite set.

2) *L is free over* $\mathbf{Z}$.

Lemma. *If V' is a finite dimensional $\mathbf{Q}$-subspace of V_K, then $L' = L \cap V'$ is a lattice of V' (i.e. is isomorphic to $\mathbf{Z}^{\dim V'}$).*

Proof.

The positive definite quadratic form $h'_c = \tilde{h}_c|_{V'_{\mathbf{R}}}$ on $V'_{\mathbf{R}}$ is such that for all M, $\{\lambda \in L', h'_c(\lambda) \leq M\}$ is finite. Therefore L' is discrete in $V'_{\mathbf{R}}$. As $V' = \mathbf{Q} \otimes L'$, we have the lemma.

We then construct a basis of L in the following way If V has finite dimension, then the lemma shows that L is free. If not, $\dim V = \aleph_0$ (V is at most countable); let $v_1, \ldots, v_n, \ldots$ be a basis of V. Let V_n be the subspace with basis $(v_1, \ldots, v_n)$. From the lemma, $L \cap V_n$ is free for all n. Suppose that one has constructed a basis $\epsilon_1, \ldots, \epsilon_n$ of $L \cap V_n$. We shall complete this to a basis of $L \cap V_{n+1}$. The image of $L \cap V_{n+1} \to V_{n+1}/V_n \cong \mathbf{Q}$ is finitely generated hence isomorphic to $\mathbf{Z}$; let $\epsilon_{n+1} \in L \cap V_{n+1}$ be such that its image is a generator. Then $\epsilon_1, ..., \epsilon_n, ...$ is a basis of L.

Complement.

Let $r(d)$ be the number of torsion points of $A(\bar{K})$ of degree $\leq d$. Then:

a) By Northcott's theorem, $r(d)$ is finite and can be effectively bounded.

b) If A is an elliptic curve without complex multiplication, we have $r(d) \asymp d^{3/2}$, as $d \to \infty$. Indeed, one knows that the Galois group is "large" and transforms one torsion point to all other torsion points of the same order (up to a finite factor). The number of points of order δ is essentially δ^2, and the number of points of degree $\leq d$ is

$$\asymp \sum_{\delta \leq d^{1/2}} \delta^2 \asymp d^{3/2}.$$

c) If A is an elliptic curve with complex multiplication, we have $r(d) \asymp d^2$.(1) Indeed, let R be the ring of integers of the imaginary quadratic field of complex multiplication of A. If $\underline{a}$ is a non-zero ideal of R, let $\phi(\underline{a})$ (Euler's function) be the number of invertible elements of $R/\underline{a}$. Then one has

$$r(d) \asymp \sum_{\underline{a} \subset R, \phi(\underline{a}) \leq d} \phi(\underline{a}).$$

It can be shown that the Dirichlet series $\sum_{\underline{a} \neq 0} \phi(\underline{a})^{-s}$ can be continued to an analytic function for $\mathrm{Re}(s) \geq 0$ with a simple pole at $s = 1$. One deduces using the Wiener-Ikehara Tauberian theorem that $\sum_{\phi(\underline{a}) \leq d} \phi(\underline{a}) \sim Cd^2$.

d) For an abelian variety of arbitrary dimension n, the order of magnitude of $r(d)$ is not easy to determine, but one can at least prove a bound $r(d) \ll d^N$, where N depends only on n.

3.10. Back to §2.11 (c algebraically equivalent to zero.)

In the next two §§, we prove two previously stated results about the elements of the Picard group.

---

(1) This was stated as a conjecture in the French text; the proof sketched below is due to R. Odoni.

Theorem. *Let V be a non-singular projective variety over K, $c \in \mathrm{Pic}^\circ(V)$, and c' an element of* $\mathrm{Pic}(V)$ *which is ample. Then*

$$h_c \leq O(h_{c'}^{1/2}) + O(1), \text{ on } V(\bar{K}).$$

Proof.

1). *The case where $V = A$ is an abelian variety.*

We may then identify c with a point of the dual A'. We may take the normalised heights $\tilde{h}_c$, $\tilde{h}_{c'}$ in place of h_c and $h_{c'}$. Further, we may suppose that c' is symmetric: $c' = [-1]^*c'$.

Associated to $c' \in \mathrm{Pic}(A)$ is a homomorphism $\phi_{c'} : A \to A'$ defined by $\phi_{c'}(a) = T_a c' - c'$. As c' is ample, $\phi_{c'}$ is an isogeny, and in particular is surjective. There is $a \in A(\bar{K})$ such that $c = \phi_{c'}(a)$ and we have (cf. §3.4):

$$\tilde{h}_c(x) = \tilde{h}_P(x, c) = \tilde{h}_P(x, \phi_{c'}(a)) = -B_{c'}(x, a),$$

where

$$B_{c'}(x, a) = \tilde{h}_{c'}(x + a) - \tilde{h}_{c'}(x) - \tilde{h}_{c'}(a).$$

But $\tilde{h}_{c'}(x)$ is a quadratic form, and we have seen that $\tilde{h}_{c'}(x) \geq 0$ as c' is ample and symmetric. The Cauchy-Schwarz inequality implies

$$|B_{c'}(x, a)| \leq B_{c'}(x, x)^{1/2} B_{c'}(a, a)^{1/2} \ll \tilde{h}_{c'}(x)^{1/2},$$

hence the result in this case.

2). *General case.*

Let A be the Albanese variety of V, and let $\phi : V \to A$ be the canonical morphism of V into A (relative to a given base point of V). It is known that if $c \in \mathrm{Pic}^\circ(V)$, then there is $c_A \in \mathrm{Pic}^\circ(A) = A'$ with $c = \phi^* c_A$. Then $h_c = h_{c_A} \circ \phi + O(1)$.

Let $c''_A \in \mathrm{Pic}(A)$ be ample, and let $c'' = \phi^* c''_A \in \mathrm{Pic}(V)$. Then

$$h_{c_A} = O(1 + h_{c''_A}^{1/2}),$$

therefore

$$h_c = O(1 + h_{c''}^{1/2}).$$

As c' is ample, we have

$$|h_{c''}| = O(1 + h_{c'}),$$

whence the result.

3.11. Back to §2.9 (torsion c).

Theorem. *Let V be a non-singular projective variety over a number field K. Let $c \in \mathrm{Pic}(V)$. If $|h_c| = O(1)$ on $V(\bar{K})$, then c is an element of finite order in $\mathrm{Pic}(V)$.*

The proof will show that there is an integer $d_V = d$, depending only on V, such that if $|h_c| = O(1)$ on the points $P \in V(\bar{K})$ of degree $\leq d$ over K, then c is torsion. For an elliptic curve, $d = 2$ suffices.

Proof.

1) *Special case when V is a curve.*

We have an exact sequence

$$0 \longrightarrow J(K) \longrightarrow \mathrm{Pic}(V) \xrightarrow{\text{degree}} \mathbf{Z}$$

where J is the Jacobian.

We show first that $\deg c = 0$. If one had $\deg c > 0$, c would be ample. We choose $d = d_V$ such that V has infinitely many points of degree $\leq d$ over K (just write the curve as a ramified covering of degree d of $\mathbf{P}_1$). From Northcott's theorem (§2.4), $h_c \to \infty$ on this set of points, thus is unbounded. The same argument applied to $-c$ shows that $\deg c = 0$.

Therefore, we have $c \in J$. The Jacobian is principally polarised, and so is equal to its dual: $J = J'$, canonically. We select a rational point of V (by making an extension of K of bounded degree); we have then a map $\phi : V \to J$ and the height h_c on V can be chosen to be $h_c = \tilde{h}_{c,J} \circ \phi$. Suppose that c is not of finite order on J (in particular, the genus is ≥ 1). By §3.4, we have

$$\tilde{h}_{c,J}(x) = -B(x, c),$$

where B is the bilinear form associated to a strictly positive quadratic form on $J(\bar{K})$ modulo torsion. As this bilinear form is non-zero, there is a point x of the Jacobian (for example $x = c$) such that $B(x, c) \neq 0$. There is then a sequence of points x_n of $J(K)$ (for example, $x_n = nc$) such that $\tilde{h}_{c,J}(x_n) \to \infty$.

Let g be the genus of the curve. It is well known that the morphism

$$\phi_g : V^g = V \times \ldots \times V \to J$$

$$(v_1, \ldots, v_g) \mapsto \sum_i \phi(v_i)$$

is surjective. We use the following simple lemma.

Lemma. *Let $f : X \to Y$ be a surjective morphism of algebraic varieties over K. There is an integer d_f such that for all fields K' with $\bar{K} \supset K' \supset K$ any point $P \in Y(K')$ can be lifted to X over an extension of K' of degree $\leq d_f$.*

[This is clear if X is finite over Y, by taking d to be the degree of the covering. We reduce to this case by *dévissage*. By induction, Y may be assumed to be irreducible and of dimension n. We then select in X a subvariety X' of dimension n for which the morphism $X' \to Y$ is surjective and finite outside a subvariety Y' of Y of dimension $< n$. On Y', one uses the induction hypothesis.]

We apply this lemma to $\phi_g : V^g \to J$. There is d_1 such that all points of $J(K)$ lift to points of degree $\leq d_1$. Let $y_n^{(1)}, ..., y_n^{(g)}$ be in $V(K_n)$ with $[K_n : K] \leq d_1$ such that

$$\sum_{j=1}^{g} \phi(y_n^{(j)}) = x_n.$$

As the height is additive (it equals $-B(x, c)$) we have

$$\tilde{h}_{c,J}(x_n) = \sum_{j=1}^{g} h_c(y_n^{(j)}).$$

But $\tilde{h}_{c,J}(x_n) \to \infty$, therefore the $h_c(y_n^{(j)})$ are unbounded.

2) *General case.*

Let V be a non-singular projective variety of any dimension. We have a homomorphism (see §3.2) $\mathrm{Pic}(V) \to \mathrm{NS}(V) \cong \mathbf{Z}^\delta \oplus$ (finite group). We first show that c has trivial image in $\mathbf{Z}^\delta$, i.e. is homologous to 0, up to torsion.

There are a finite number of curves $C_i \subset V$ (in fact, δ curves suffice) defined over $\bar{K}$, such that if $c \in \mathrm{Pic}(V)$ induces an element of degree 0 on each C_i, then c is homologically equivalent to zero, up to torsion. In terms of divisors, the cohomology class, with coefficients in $\mathbf{Q}$, of a divisor D is zero if its intersection number with each C_i is zero.

As we have assumed that h_c is bounded, we may apply the above: on each curve h_c is bounded, therefore c is homologically equivalent to zero, up to torsion. Replacing c by a non-zero multiple, we may then suppose that $c \in \mathrm{Pic}^\circ(V)$.

We now select a curve C on V which is a non-singular plane section (that is, if $\dim V = n$, take a section by a linear variety of codimension $n-1$

in general position). The restriction map $\mathrm{Pic}^{\circ}(V) \to \mathrm{Pic}^{\circ}(C)$ is known to be injective (it is enough to know that its kernel is finite, which is easy). From 1), c is of finite order in $\mathrm{Pic}^{\circ}(C)$, thus is of finite order in $\mathrm{Pic}^{\circ}(V)$.

4. THE MORDELL-WEIL THEOREM

4.1. Hermite's finiteness theorem.

Theorem. *There are only finitely many number fields K (up to isomorphism) of given discriminant d_K (and of given degree $n_K = [K : \mathbf{Q}]$).*

If $n_K \geq 2$, one has $|d_K| \geq (\sqrt{3})^{n_K}$ (there are better results, for example, $n_K > 20 \Rightarrow |d_K| > 10^{n_K}$ (Odlyzko)), thus if the discriminant is bounded, the degree is also bounded.

Sketch of proof.
Assume for simplicity that K is totally real (the general case is similar, see e.g. [L6]). Let $n = n_K$. The integers O_K of the field K form a lattice in $\mathbf{R}^n$. Applying Minkowski's theorem, there is a constant C (depending only on the discriminant) and a non-zero element $x \in O_K$ such that if $x_1, ..., x_n$ are the conjugates of x, then one has

$$|x_1| \leq C, |x_2| \leq 1/3, ..., |x_n| \leq 1/3.$$

The integer x generates the field K and the coefficients of its minimum polynomial are bounded by a function of the discriminant. Therefore, there are only finitely many such x, and such fields K, up to isomorphism.

Variant. *Let K be a number field, S a finite set of places of K, and n an integer ≥ 1. Then there are only finitely many extensions L of K (up to isomorphism) of degree n which are unramified outside S.*

This variant is deduced from Hermite's theorem in the following way. We may first reduce to the case $K = \mathbf{Q}$. It remains to bound the discriminant of L in terms of S and n only. Such a bound has been given by Hensel:

$$|d_{L/\mathbf{Q}}| = \prod_{p \in S} p^{\delta_p},$$

with

$$\delta_p \leq n - 1 + n \log(n)/\log(p),$$

see e.g. §1 of [Se1].

For the application below, the following particular case of Hermite's theorem is sufficient (and can be proved directly from the unit theorem): *There are only finitely many cyclic extensions of K of given degree which are unramified outside S.*

4.2. The Chevalley-Weil theorem.

Theorem. (Chevalley-Weil, [CW]). *Let K be a number field. Let $\pi : Y \to X$ be a finite étale morphism of degree d of projective varieties over K. There is a finite set of places S of K such that if K' is a finite extension of K and if $P \in X(K')$, then there exists $K'' \supset K'$ and $Q \in Y(K'')$ such that $\pi(Q) = P$, $[K'' : K'] \leq d$, and the extension K''/K' is unramified outside the places $S_{K'}$ of K' above S.*

There is an analogous statement for integral points without the projectivity hypothesis (a non-projective case: $\mathbf{G}_m \to \mathbf{G}_m$, $x \mapsto x^2$; the square roots of units are unramified except at 2).

Proof.

Let O be the ring of integers of K; if S is a finite set of non-archimedean places of K, one can view S as a subset of Spec O; let O_S be the ring of S-integers:

$$\text{Spec } O_S = \text{Spec } O - S, \quad \text{and} \quad \bigcup_S O_S = K.$$

From [G, IV,§8], all algebraic constructions over K can be made over some O_S for S sufficiently large. Hence, there are finite étale morphisms $\pi_S : Y_S \to X_S$ of projective O_S-schemes which induce π by the base change $O_S \to K$. Select such a suitable S.

Let O' be the ring of integers of K'. From the viewpoint of schemes, $P \in X(K')$ corresponds to a morphism Spec$(K') \to X$ which, as X is projective, provides a morphism $P : \text{Spec}(O'_{S_{K'}}) \to X_S$, where $S_{K'}$ is the set of places of K' lying over S. Base changing π_S by the morphism P, we obtain a scheme Λ, finite and étale over Spec$(O'_{S_{K'}})$, which provides $Q \in Y(K'')$:

$$\begin{array}{ccc} \Lambda & \xrightarrow{Q} & Y_S \\ \downarrow & & \downarrow \pi_S \\ \text{Spec}\, O'_{S_{K'}} & \xrightarrow{P} & X_S \end{array}$$

The theorem can be refined and stated for an arbitrary field: the discriminant of the extension K''/K' divides a fixed element of K (see, e.g. [L6], Chapter 2, Theorem 8.1.). Over a number field, these are equivalent, because of Hermite's theorem (§4.1).

4.3. Mordell-Weil theorem.

Theorem. (Weak Mordell-Weil theorem.) *Let K be a number field, A an abelian variety over K, and $\Gamma = A(K)$. Then $\Gamma/n\Gamma$ is finite for all integers $n \geq 1$.*

Proof. (Under the hypothesis that the n-torsion points of A be K-rational.)

Let A_n be the kernel of the map $\pi : x \mapsto nx$ on A. From the Chevalley-Weil theorem (§4.2), applied to the covering π, if $P \in A(K)$ then there is $Q \in A(K')$, where $[K' : K] \leq d(= n^{2\dim A})$ such that $nQ = P$, and K'/K is unramified outside a finite number of places. By Hermite's theorem (§4.1), the corresponding fields K' are finite in number; let L be a finite Galois extension of K containing all the K'. Let G be the Galois group of L/K. We shall embed $\Gamma/n\Gamma$ in $\mathrm{Hom}(G, A_n)$:

If $P \in \Gamma$, there is $Q \in A(L)$, with $P = nQ$. Let $s \in G$; we have

$$P = sP = nsQ,$$

therefore

$$n(sQ - Q) = 0,$$

whence

$$sQ - Q \in A_n.$$

As Q is unique, up to addition by an element of A_n, and as the elements of A_n are rational over K (by hypothesis), $sQ - Q$ is independent of the choice of Q. For $P \in \Gamma$ and $s \in G$ we put $\epsilon(P)(s) = sQ - Q$. For $s, t \in G$, we have

$$\begin{aligned} stQ - Q &= s(tQ - Q) + sQ - Q \\ &= (tQ - Q) + (sQ - Q), \end{aligned}$$

therefore, $\epsilon(P) \in \mathrm{Hom}(G, A_n)$. If $P = nP'$ with $P' \in \Gamma$, we have $\epsilon(P) = 0$, thus ϵ factors through a homomorphism $\Gamma/n\Gamma \to \mathrm{Hom}(G, A_n)$. Finally, if $sQ - Q = 0$ for all s then $Q \in \Gamma$ thus $P = nQ \in n\Gamma$, whence we have an injective homomorphism, $\epsilon : \Gamma/n\Gamma \to \mathrm{Hom}(G, A_n)$. As $\mathrm{Hom}(G, A_n)$ is finite, we deduce the finiteness of $\Gamma/n\Gamma$.

[*Cohomological variant.* By applying Galois cohomology to the exact sequence

$$0 \longrightarrow A_n \longrightarrow A(\bar{K}) \xrightarrow{\ n\ } A(\bar{K}) \longrightarrow 0$$

one gets an injection $\epsilon : \Gamma/n\Gamma \to H^1(K, A_n)$, where
$H^1(K, A_n) = H^1(\mathrm{Gal}(\bar{K}/K), A_n)$. If S is a finite set of places of K, let

$H^1_S(K, A_n)$ denote the subgroup of $H^1(K, A_n)$ made up of the elements which are "unramified outside S". One then proves that:

a) $H^1_S(K, A_n)$ is finite (by Hermite's theorem);

b) $\epsilon(\Gamma/n\Gamma)$ is contained in $H^1_S(K, A_n)$ if S is large enough (e.g. if S contains all places dividing n, and all places where A has bad reduction.

This shows that $\Gamma/n\Gamma$ is finite.

In the proof above, we have assumed that the points of A_n are rational over K so that $H^1(K, A_n)$ can then be identified with $\mathrm{Hom}(\mathrm{Gal}(\bar{K}/K), A_n)$ and one can dispense with the cohomological language.]

Non-abelian generalisation. The above theorem could be reformulated by saying that, if $\pi : B \to A$ is an isogeny of abelian varieties, then $\mathrm{Coker}\{\pi : B(K) \to A(K)\}$ is finite. There is a similar statement for algebraic groups (not necessarily commutative), with $A(K)$ and $B(K)$ replaced by suitable "S-arithmetic" groups. The proof is basically the same. See e.g. Borel [Bor] and Honda [Ho].

We can now prove:

Mordell-Weil Theorem. *Let K be a number field and A an abelian variety over K. Then the group $A(K)$ is finitely generated.*

Proof.

Select an integer $n \geq 2$. Let $K' = K(A_n)$; we have just proved that $A(K')/nA(K')$ is finite. But, by §3.9, $A(K')$ is a direct sum of a free abelian group and a finite group. Hence $A(K')$ is finitely generated and the same is true for $A(K)$, since $A(K)$ is a subgroup of $A(K')$.

Effectivity. The group $\Gamma/n\Gamma$ is embedded effectively in a finite group, whence one has an effective bound for $|\Gamma/n\Gamma|$, and also for the rank of Γ. But there is no known effective method for characterising the elements of $\Gamma/n\Gamma$ in the image nor for finding *a priori* a set of representatives $\gamma_i \in \Gamma$ of $\Gamma/n\Gamma$. It is this which makes the search for rational points on elliptic curves an art but not a science.

Remark. The Mordell-Weil theorem has been extended by Néron [Ne] to the case where K is only assumed to be finitely generated over the prime field ($\mathbf{Q}$ or $\mathbf{F}_p$). See e.g. Lang [L6], where this statement is reduced to the standard one by means of the Lang-Néron theorem (if K is a function field over a ground field k, A an abelian variety over K, and $\tau : B \to A$ a "K/k-trace" of A, then $A(K)/\tau B(K)$ is a finitely generated group).

4.4 The classical descent.

Let Γ be a $\mathbf{Z}$-module, and let V be a $\mathbf{Q}$-vector space containing $\Gamma/\Gamma_{\text{tors}}$. We suppose given a real valued positive quadratic form on V, written $x.x = |x|^2$. Let $n \geq 2$ be an integer, and let $\gamma_i \in \Gamma$ be representatives of $\Gamma/n\Gamma$. Let C be a real number with $C > 0$. Assume that $|\gamma_i| \leq C$ for all i.

Descent Lemma. *The group Γ is generated by the elements $\gamma \in \Gamma$ such that $|\gamma| \leq 2C$.*

Proof.

For each real number $D > 0$, we denote by Γ_D the subgroup of Γ generated by the elements $\gamma \in \Gamma$ with $|\gamma| \leq D$. We wish to show that $\Gamma_{2C} = \Gamma$; as $\bigcup_{m\geq 1} \Gamma_{mC} = \Gamma$, it suffices to check by induction on m that

$$\Gamma_{mC} \subset \Gamma_{(m-1)C}, \quad \text{for} \quad m \geq 3.$$

Let $y \in \Gamma_{mC}$; there is i and $x \in \Gamma$ with

$$y = nx + \gamma_i.$$

Then

$$n|x| = |y - \gamma_i| \leq |y| + |\gamma_i|,$$

thus

$$\begin{aligned} |x| &\leq \frac{1}{2}(|y| + |\gamma_i|) \\ &\leq \frac{(m+1)}{2}C. \end{aligned}$$

As $m \geq 3$, we have, $(m+1)/2C \leq (m-1)C$, thus $x \in \Gamma_{(m-1)C}$.

Proof of the Mordell-Weil theorem via the descent lemma.

The lemma is applied to $\Gamma = A(K)$, with $|x|^2$ being the normalised height associated to some ample symmetric divisor class. By the weak Mordell-Weil theorem (§4.3), $\Gamma/n\Gamma$ is finite, hence C exists; since the elements of $A(K)$ with $|\gamma| \leq C$ are finite in number (Northcott's theorem, §2.4), this shows that $A(K)$ is finitely generated.

4.5. The number of points of bounded height on an abelian variety.

Assume K is a number field, and choose a projective embedding $\phi : A \to \mathbf{P}_n$, whence an associated logarithmic height $h = \log(H \circ \phi)$. Let $N(X)$ be the number of $\gamma \in \Gamma$ such that $h(\gamma) \leq X$.

Theorem. (Néron, [N4]). *We have*

$$N(X) = (w/\sqrt{D})c_\rho X^{\rho/2} + O(X^{(\rho-1)/2}), \text{ as } X \to \infty,$$

where

$\Gamma = \Gamma_{\text{tors}} \oplus \mathbf{Z}^\rho$,
w *is the order of* Γ_{tors},
ρ *is the rank of* $\Gamma/\Gamma_{\text{tors}}$,
c_ρ *is the volume of the unit ball in* $\mathbf{R}^\rho$: $c_\rho = \pi^{\rho/2}/\Gamma(1+\rho/2)$,
$D = \det(B(\gamma_i, \gamma_j)/2)$, *where* $\gamma_1, ..., \gamma_\rho$ *is a basis of* $\Gamma/\Gamma_{\text{tors}}$, *and* B *is the bilinear form attached to the normalised height* $\tilde{h}$:

$$B(x,y) = \tilde{h}(x+y) - \tilde{h}(x) - \tilde{h}(y).$$

[If the element of the group $\text{Pic}(A)$ attached to ϕ is symmetric, we may replace $O(X^{(\rho-1)/2})$ by $O(X^{\rho/2-\rho/(\rho+1)})$.]

Corollary. $N(X) \sim cX^{\rho/2}$, *as* $X \to \infty$, *with* $c = c_\rho w/\sqrt{D}$.

Proof of theorem.

For $\rho = 0$ we have $N(X) = w$ for X sufficiently large. We now suppose that $\rho \geq 1$.

It suffices to prove the same estimate for the number $\tilde{N}(X)$ of $\gamma \in \Gamma$ for which $\tilde{h}(\gamma) \leq X$. Indeed, h and $\tilde{h}$ differ by a bounded quantity: $|h - \tilde{h}| \leq a$, therefore

$$\tilde{N}(X-a) \leq N(X) \leq \tilde{N}(X+a).$$

Thus if we prove the estimate for $\tilde{N}$, we have

$$\tilde{N}(X+a) \leq c(X+a)^{\rho/2} + O(X^{(\rho-1)/2})$$

$$= cX^{\rho/2} + O(X^{\rho/2-1}) + O(X^{(\rho-1)/2}) \leq cX^{\rho/2} + O(X^{(\rho-1)/2}),$$

whence the estimate for $N(X)$.

It is enough to count the points modulo torsion, provided that one finally multiplies by w.

Identifying $\Gamma/\Gamma_{\text{tors}}$ with $\mathbf{Z}^\rho$ by means of the basis $\gamma_1, ..., \gamma_\rho$, we are led to count the $(n_1, ..., n_\rho) \in \mathbf{Z}^\rho$ with

$$\tilde{h}(\sum_{i=1}^{\rho} n_i \gamma_i) \leq X.$$

But $\tilde{h}$ is a positive definite quadratic form plus a linear term in the γ_i. Let $\tilde{N}_{\text{quad}}$ be the number of γ for which the quadratic part $\tilde{h}_{\text{quad}}$ of $\tilde{h}$ is $\leq X$. Then

$$\tilde{N}_{\text{quad}}(X - bX^{1/2}) \leq \tilde{N}(X) \leq \tilde{N}_{\text{quad}}(X + bX^{1/2}),$$

for a certain constant b. It is then enough to study $\tilde{N}_{\text{quad}}(X)$.

This means one has to estimate the number of integral points in the interior of the ellipsoid $\tilde{h}_{\text{quad}} \leq X$. This number is equal to the volume $cX^{\rho/2}$ of the ellipsoid $\tilde{h}_{\text{quad}} \leq X$ up to some $O(X^\alpha)$ error term. Elementary methods give $\alpha = \rho/2 - 1/2$, which is enough for what we want (by [La] and [CN], one can take $\alpha = \rho/2 - \rho/(\rho+1)$, and even better results are known when $\rho = 2$).

The constant c is the volume of the ellipsoid for $X = 1$. If the quadratic form is $\sum x_i^2$, we find c_ρ. If not, we change the basis and the volume change appears as the determinant of a matrix, whence the factor $\sqrt{D}$.

4.6. Explicit form of the weak Mordell-Weil theorem.

As in §4.3, let n be an integer ≥ 1 and A_n the kernel of $[n] : A \to A$. Let μ_n be the group of roots of unity, and d the dimension of A.

Assume:

(1) $A_n \subset A(K)$;

(2) $\mu_n \subset K$.

[*Exercise.* Prove that (1) $\Rightarrow$ (2), if $d \geq 1$.]

The covering $[n] : A \to A$ is abelian and unramified with Galois group A_n, thus of type $(n, ..., n)$ and of degree n^{2d}. Let K_A be the function field of A. Using Kummer theory, one sees that:

The function field extension of K_A corresponding to the covering $[n]$: $A \to A$ is the composite of the extensions $K_A(\sqrt[n]{\phi})$, where ϕ runs through the elements of K_A^* having the following two properties:

(i) The divisor (ϕ) is divisible by n;

(ii) ϕ is defined at 0 and $\phi(0) = 1$.

(Condition (ii) expresses the fact that 0 splits in the covering.)

Suppose that one has explicitly the functions $\phi_1, ..., \phi_{2d}$ describing the covering $[n] : A \to A$.

We defined in §4.3 an embedding

$$\Gamma/n\Gamma \hookrightarrow \operatorname{Hom}(\operatorname{Gal}(\bar{K}/K), A_n).$$

As A_n is an abelian group of type $(n, ..., n)$,

$$\mathrm{Hom}(\mathrm{Gal}(\bar{K}/K), A_n) \cong \mathrm{Hom}(\mathrm{Gal}(\bar{K}/K), \mathbf{Z}/n\mathbf{Z} \times ... \times \mathbf{Z}/n\mathbf{Z})$$

$$\cong \mathrm{Hom}(\mathrm{Gal}(\bar{K}/K), \mu_n) \times ... \times \mathrm{Hom}(\mathrm{Gal}(\bar{K}/K), \mu_n).$$

Using Kummer theory again, we have $\mathrm{Hom}(\mathrm{Gal}(\bar{K}/K), \mu_n) \cong K^*/K^{*n}$, hence an injection

$$\Gamma/n\Gamma \hookrightarrow K^*/K^{*n} \times ... \times K^*/K^{*n},$$

which can be shown to be:

$$\gamma \in \Gamma \mapsto (\phi_1(\gamma), ..., \phi_{2d}(\gamma)).$$

[Here $\phi_j(\gamma)$ is the value (mod K^{*n}) of ϕ_j at γ in case the divisor of ϕ_j does not go through γ; in case it does, one just replaces ϕ_j by $\phi_j f^n$, with a suitable $f \in K_A^*$, so that this does not happen.]

From this point of view, the proof of the weak Mordell-Weil theorem reduced to saying that *the image of* $\gamma \mapsto \phi_j(\gamma)$ *is a finite subgroup of* K^*/K^{*n} *for each* j.

Example: elliptic curves.

We take $n = 2$, $d = 1$ and write the equation of A as

$$y^2 = 4(x - e_1)(x - e_2)(x - e_3),$$

with e_1, e_2, e_3 distinct elements of K, which we may assume to be integers.

The divisor $D_1 = (P_1) - (0)$ where $P_1 = (e_1, 0)$ and $0 = (\infty, \infty)$, is such that $2D_1 = (\phi_1)$ with $\phi_1 = (x - e_1) = (x - e_2)/(x - e_3)\times$ square. Define similarly $\phi_2 = (x - e_2)$. The construction above gives an injective homomorphism

$$\epsilon : \Gamma/2\Gamma \to K^*/K^{*2} \times K^*/K^{*2},$$

$$\gamma = (x, y) \mapsto (\phi_1(\gamma), \phi_2(\gamma)).$$

(*Exercise.* Check that ϵ is a homomorphism, and that it is injective, using explicit addition formulae.)

The finiteness of the image of ϵ follows from the more precise statement:

Lemma. *If $\wp$ is a prime ideal of K not dividing $(e_1 - e_2)(e_2 - e_3)(e_3 - e_1)$, then*

$$v_\wp(x - e_1) \equiv 0 \pmod 2 \text{ and } v_\wp(x - e_2) \equiv 0 \pmod 2,$$

for any $(x, y) \in \Gamma$.

Proof.

Put $m_i = v_\wp(x - e_i)$, $i = 1, 2, 3$. Since $(x - e_1)(x - e_2)(x - e_3)$ is a square, we have $m_1 + m_2 + m_3 \equiv 0 \pmod 2$. To prove that $m_i \equiv 0 \pmod 2$ for all i we distinguish two cases:

a) x is not $\wp$-integral. If $v_\wp(x) = -m$, then $m_i = -m$ for $i = 1, 2, 3$, hence $3m \equiv 0 \pmod 2$ and $m_i \equiv 0 \pmod 2$.

b) If x is $\wp$-integral, at most one of the m_i can be $\neq 0$ (otherwise, one of the $e_i - e_j$ would be divisible by $\wp$). Hence the result by using again $m_1 + m_2 + m_3 \equiv 0 \pmod 2$.

Exercise.

Let E be an elliptic curve over $\mathbf{C}$ given in standard Weierstrass form

$$\wp'(u)^2 = 4\wp(u)^3 - g_2\wp(u) - g_3 = 4(\wp(u) - e_1)(\wp(u) - e_2)(\wp(u) - e_3),$$

and denote by ω_1, ω_2 a pair of fundamental periods, so that

$$e_1 = \wp(\omega_1/2), e_2 = \wp(\omega_2/2), e_3 = \wp(\omega_1/2 + \omega_2/2).$$

Let $y_i(u)$, $i = 1, 2, 3$, be defined by:

$$y_i(u) = \sqrt{\wp(u) - e_i}, y_i = \frac{1}{u} + \ldots.$$

The field generated by y_1, y_2, y_3 over $\mathbf{C}(\wp(u), \wp'(u))$ is the function field of 2-division of E, i.e. $\mathbf{C}(\wp(u/2), \wp'(u/2))$. Check this by using the relations (cf. [Wh]):

$$\wp(u/2) = \wp(u) + y_1y_2 + y_2y_3 + y_3y_1,$$

$$\frac{1}{2}\wp'(u/2) = \wp'(u) - (y_1^2y_2 + y_1^2y_3 + y_2^2y_1 + y_2^2y_3 + y_3^2y_1 + y_3^2y_2),$$

and

$$y_i = -\wp'(u)^{-1}(\wp(u/2) - \wp(\omega_i/4))(\wp(u/2) - \wp(\omega_j/2 + \omega_i/4)),$$

for $i, j \in \{1, 2\}, i \neq j$.

5. MORDELL'S CONJECTURE

The conjecture is the following [it has now been proved by Faltings, see [F,FW,Sz]].

(M) *If C is a curve over a number field K, and if the genus of C is ≥ 2, then the set $C(K)$ of points of C rational over K is finite.*

It is no restriction to assume that the curve is projective and non-singular, for this changes only a finite number of points.

We shall give three partial results on this conjecture. The first two (Chabauty [Ch] and Demjanenko and Manin [Ma1]) prove the finiteness of $C(K)$ under certain hypotheses, the third (Mumford [Mu1]) shows that $C(K)$ is "sparse".

5.1. Chabauty's theorem.

We select a K-rational point P_0 on the curve, and map C into its Jacobian by

$$\phi : C \to J = \operatorname{Jac}(C)$$

$$P \mapsto \operatorname{Cl}((P) - (P_0)).$$

This is an embedding, which allows us to identify C with its image:

$$C \subset J, \quad C(K) = C \cap J(K).$$

By the Mordell-Weil theorem, $J(K)$ is a finitely generated abelian group. It is natural to make the following conjecture:

(M′) *If A is an abelian variety over a field L of characteristic zero,* $\dim A \geq 2$, *C a curve contained in A which generates A, and Γ a finitely generated subgroup of $A(L)$, then $C(L) \cap \Gamma$ is finite.*

One says that C generates A if the differences $P - Q$ $(P, Q \in C)$ generate the group A. Alternatively, one may suppose after translation that $0 \in C$, and then ask that C generates A in the usual sense. This amounts to the homomorphism $J \to A$ being surjective (in the geometric sense, i.e. over an algebraic closure).

It is clear that (M′) $\Rightarrow$ (M). Conversely,

Theorem. (M) $\Rightarrow$ (M′).

The proof is the following. As everything is defined over a field finitely generated over $\mathbf{Q}$, we may suppose that L is finitely generated over $\mathbf{Q}$. A theorem of Néron (see §11.1, [N1]) gives a specialisation of L to a number field K and an injective homomorphism $A(L) \to A_K(K)$, where A_K is the corresponding specialised variety of A. One is then reduced to the case of a number field. The Jacobian J of C projects onto A, therefore the genus of C is ≥ 2, and from (M), $C(L)$ is finite.

Theorem. (Chabauty [Ch]). *As in conjecture (M′), let L be a field of characteristic zero, C a generating curve in A, and Γ a finitely generated subgroup of $A(L)$. Suppose that* rank $\Gamma < \dim A$. *Then, $C(L) \cap \Gamma$ is finite.*

Consequently, conjecture (M) is true for the pair (C, K) if rank $J(K) <$ genus of C.

The proof uses the following well known result:

Lemma. *If $\omega \neq 0$ is a differential form of the first kind on A, then the restriction of ω to the curve C is $\neq 0$.*

Proof of lemma.

The injection of C in A factors through the Jacobian of the curve,

$$\begin{array}{ccc} C & \xrightarrow{\phi} & J \\ \downarrow & \swarrow & \\ A & & \end{array}$$

and by hypothesis, the map $J \to A$ is surjective; therefore the tangent map is surjective, and the differential forms of A embed in those of J. As ϕ gives an isomorphism between the space $H^0(J, \Omega^1_J)$ of differential forms of J and the space $H^0(C, \Omega^1_C)$ of those on C, the lemma follows.

[This lemma can also be proved by noting that the product map $C \times \ldots \times C \to A$ is surjective, if sufficiently many factors are taken; if ω_C is the restriction of ω to C, then the inverse image of ω is $\omega_C \otimes 1 + \ldots + 1 \otimes \omega_C$.]

Proof of Chabauty's theorem.

1). First reduction. It suffices to prove the theorem when L is finitely generated over $\mathbf{Q}$. Indeed, the group Γ is finitely generated and everything is defined over a finitely generated extension of $\mathbf{Q}$.

2). We now embed L in a finite extension of $\mathbf{Q}_p$ (this is possible, see the appendix at the end of this section).

3) We may thus assume that L is a finite extension of $\mathbf{Q}_p$; hence L is a locally compact field and $A(L)$ is a compact analytic group over L of dimension $d = \dim A$. Such a group is locally isomorphic to its Lie algebra: there is an open subgroup U of $A(L)$ isomorphic, as a Lie group, to $O_L \times \ldots \times O_L$ (d factors), where O_L is the ring of integers of L.

[This can be proved as in the complex case by taking a basis of the differential forms of the first kind, and integrating them locally; assuming that $0 \in C$, one obtains the coordinates $x_i = \int_0^P \omega_i$ which give a local isomorphism between $A(L)$ and its Lie algebra.

Another method is to use the logarithm map

$$\log : A(L) \to \mathrm{Lie}(A),$$

which is everywhere defined (since $A(L)$ is compact) and gives a local isomorphism (cf. Bourbaki, Lie III, §7.6).]

4). Let us now show that $C \cap \bar{\Gamma}$ (where $\bar{\Gamma}$ is the closure of Γ in $A(L)$) is finite. Suppose that $C \cap \bar{\Gamma}$ is infinite; as $A(L)$ is compact, we can extract a sequence of distinct points P_i of $C \cap \bar{\Gamma}$, such that $P_i \to P_0$ where $P_0 \in C \cap \bar{\Gamma}$.

We may assume that $P_0 = 0$, by translation, that P_i belongs to the open subgroup U chosen above, and that $\bar{\Gamma} \cap U$ is free of rank $d' < d$. Changing coordinates, we then have that all $\gamma = (x_1, \ldots, x_d) \in U \cap \bar{\Gamma}$ satisfy $x_1 = 0$. Thus in d-dimensional space, the analytic curve C intersects the hyperplane $x_1 = 0$ in infinitely many points; in other words, x_1 is a function on the curve having infinitely many zeros $x_1(P_i) = 0$, which accumulate at 0. The principle of isolated zeros for p-adic analytic functions, gives $x_1 = 0$ in a neighbourhood of 0 on C.

Let $\omega_1, \ldots, \omega_d$ be a base of the differential forms corresponding to the system of coordinates $x_1, \ldots, x_d$: $dx_i = \omega_i$ in a neighbourhood of 0. The differential form ω_1, restricted to C has at most $2g - 2$ zeros, therefore $\omega_1|_C = 0$, which contradicts the lemma.

Chabauty proved this result in 1941. Previously, Skolem had already used p-adic methods to study products of multiplicative groups. For an exposition of Skolem's method, see e.g. [BSh], Chap.IV, §6.

Remark. Chabauty's method can sometimes be used to give an effective (and small) bound for the number of rational points on a curve, cf. [Col].

Appendix. Embedding a field finitely generated over $\mathbf{Q}$ *in* $\mathbf{Q}_p$.

Theorem. *Let* L *be a field finitely generated over* $\mathbf{Q}$.

a) *For every prime* p *there is an embedding of* L *in some finite extension of* $\mathbf{Q}_p$.

b) *For infinitely many primes* p *there is an embedding of* L *in* $\mathbf{Q}_p$.

Proof. a) Let $\bar{\mathbf{Q}}_p$ be an algebraic closure of $\mathbf{Q}_p$. As the transcendence degree of $\bar{\mathbf{Q}}_p$ over $\mathbf{Q}$ is infinite, L is embedded in $\bar{\mathbf{Q}}_p$. The compositum $L.\mathbf{Q}_p$ is then an algebraic extension of $\mathbf{Q}_p$ (since it is contained in $\bar{\mathbf{Q}}_p$), and finitely generated; it is therefore a finite extension of $\mathbf{Q}_p$.

Let L_0 be an algebraic closure of $\mathbf{Q}$ in L (that is, $L_0 = L \cap \bar{\mathbf{Q}}$). We write $L = \mathbf{Q}(V) = L_0(V)$, where V is an absolutely irreducible algebraic variety over L_0; as V is defined up to birational isomorphism, we may suppose that V is affine and smooth. We wish to embed L in $\mathbf{Q}_p$; we begin by embedding L_0.

Consider the prime ideals $\wp | p$ of the number field L_0 where p does not divide the discriminant of L_0; then, L_0 embeds in $\mathbf{Q}_p$ if and only if one of the $\wp$'s has degree 1. Denote by $P(L_0)$ the set of p for which L_0 embeds in $\mathbf{Q}_p$. By Chebotarev's theorem, the density of such p's is > 0.

Proposition. *For all* $p \in P(L_0)$, *except finitely many, the field* L *embeds in* $\mathbf{Q}_p$.

Proof of proposition.

Recall that V is affine. Let $\Lambda = L_0[V]$ be its affine coordinate ring. A point $P \in V(\mathbf{Q}_p)$ defines a homomorphism $\Lambda \to \mathbf{Q}_p$, and we say that P is generic if this homomorphism is injective. We have to prove that such generic points exist.

Lemma. *If* $P_0 \in V(\mathbf{Q}_p)$ *is a smooth point, then* $V(\mathbf{Q}_p)$ *contains generic points arbitrarily near to* P_0.

Proof of the lemma.

The elements $f \in \Lambda$, $f \neq 0$, form a countable family. The equation $f = 0$ cuts out a subvariety W_f of dimension $< \dim V$. These varieties have no interior points (p-adically). We now use Baire's category theorem (or even measure theory): the complement in $V(\mathbf{Q}_p)$ of the union of the $W_f(\mathbf{Q}_p)$ is dense. If P belongs to that complement, then P is generic.

Thus to prove the proposition, it suffices to show that V has at least one smooth $\mathbf{Q}_p$-rational point. To do so, one selects a finite subset S of Spec O_{L_0} such that V comes by extension of scalars from an affine scheme

V_0 over Spec $O_{L_0} - S$. If $\wp \in$ Spec $O_{L_0} - S$, V_0 may then be reduced mod $\wp$; this gives a variety $V_\wp$ over the residue field $\kappa(\wp)$. It is known that $V_\wp$ is smooth and absolutely irreducible for almost all $\wp$. Moreover:

Lemma. *The variety $V_\wp$ has a $\kappa(\wp)$-rational point for almost all $\wp \in$ Spec (O_{L_0}).*

Proof. One may reduce to the case of curves; from a theorem of Weil, the number of points on a non-singular projective curve is $\geq q + 1 - 2g\sqrt{q}$, where $q = N(\wp)$ and g is the genus (see also the Lang-Weil theorem §13.3). Thus for q sufficiently large, it has at least one point.

Given a smooth point over $\kappa(\wp)$, Hensel's lemma provides a smooth point of V over the completion $\tilde{O}_\wp$ of the local ring of O_{L_0} at $\wp$. So the variety V has a point over the completion $L_{0,\wp}$ of L_0 at $\wp$. For the $\wp$ of degree 1, we have $L_{0,\wp} = \mathbf{Q}_p$.

5.2. The Manin-Demjanenko theorem.

Let C be a projective non-singular absolutely irreducible variety over a number field K. Let A be an abelian variety over K, and $f_1, \ldots, f_m$ morphisms from C to A defined over K.

We make two hypotheses:

First hypothesis. The f_i are independent in the following sense. Let $P_0 \in C(K)$, and let $\tilde{f}_i(x) = f_i(x) - f_i(P_0)$. Then $\tilde{f}_1, \ldots, \tilde{f}_m$ are independent in the group Morph(C, A) (i.e. $\sum n_i \tilde{f}_i = 0$, for integers n_i, $\Rightarrow n_i = 0$ for all i).

This amounts to saying that if $\sum n_i f_i$ is constant, then $n_i = 0$ for all i (hence the condition is independent of the choice of P_0). If we denote by Alb(C) the Albanese variety of C, f_i defines a homomorphism $\phi_i : \text{Alb}(C) \to A$, and the hypothesis means that $\phi_1, \ldots, \phi_m$ are independent in Hom(Alb$(C), A)$.

Second hypothesis. The Néron-Severi group NS(C) has rank 1.

If C is a curve, NS(C) = $\mathbf{Z}$ (NS(C) is the group of divisor classes modulo algebraic equivalence) and the hypothesis is satisfied.

Theorem 1. *For all $P \in C(K)$ but a finite number, the points $\tilde{f}_1(P), ..., \tilde{f}_m(P) \in A(K)$ are linearly independent over $\mathbf{Z}$.*

This has the following direct consequence:

Theorem 2.([Ma1]). *Under these hypotheses, if $m >$ rank $A(K)$, then $C(K)$ is finite.*

Theorem 2 immediately implies:

Corollary. *Let C be a curve. Assume there exists an abelian variety A over K such that*

$$\text{rank Mor}(C, A) > \text{ rank } A(K).$$

Then $C(K)$ is finite.

[The proof will be effective.]

Proof of theorem 1.

Let $H = \text{Hom}(\text{Alb}(C), A)$; this is the set of morphisms $f : C \to A$ such that $f(P_0) = 0$; H is a free $\mathbf{Z}$-module of finite rank $H \cong \mathbf{Z}^\delta$; let $H_\mathbf{R} = \mathbf{R} \otimes H$. We shall suppose $f_i(P_0) = 0$ (that is to say, $\tilde{f_i} = f_i$); thus $f_i \in H \subset H_\mathbf{R}$. Let $V_\mathbf{R} = \mathbf{R} \otimes A(K)$.

We select:

- a projective embedding of C in $\mathbf{P}_N$; it corresponds to an element $c_C \in \text{Pic}(C)$;
- a symmetric and very ample element $c_A \in \text{Pic}(A)$; this corresponds to an embedding $\psi : A \to \mathbf{P}_M$.

For $f \in H$, we have $f^*c_A \in \text{Pic}(C)$; as we have a projective embedding of C, the degree of f^*c_A is defined.

Lemma 1. *The map $f \mapsto \deg f^*c_A$ is a quadratic form with integer values on H which extends to a positive definite quadratic form on $H_\mathbf{R}$.*

We already know that the map

$$H \to \text{Pic}(C), \qquad f \mapsto f^*c_A$$

is quadratic (§3.2).

If $f \neq 0$, $\deg f^*c_A > 0$. On $H_\mathbf{Q} = \mathbf{Q} \otimes H$, this quadratic form is rational valued and positive except at 0 hence non-degenerate. When one extends the scalars to $\mathbf{R}$, it remains positive definite.

[This form can be interpreted as a canonical height. A morphism $f : C \to A$ is a point of A with values in the function field K_C of C; as we have chosen a projective embedding of C, we have a product formula (cf. §2.2), and the natural height on $A(K_C)$ is $\deg\ (f^*c_A)$; it is also the canonical height since it is quadratic.]

On $V_{\mathbf{R}}$, the canonical height attached to c_A defines a positive definite quadratic form. Hence both $H_{\mathbf{R}}$ and $V_{\mathbf{R}}$ have natural Euclidean space structures.

We denote the corresponding quadratic forms by:

$$|f|^2 = \deg\,(f^* c_A), \quad \text{if} \quad f \in H,$$

$$|a|^2 = \tilde{h}_{c_A}(a), \quad \text{if} \quad a \in A(K).$$

On the other hand, if $P \in C(K)$, P defines a morphism $f \mapsto f(P)$ of H to $A(K)$; by extension of scalars we obtain a morphism $\phi_P : H_{\mathbf{R}} \to V_{\mathbf{R}}$. We shall show that *$\phi_P$ is injective for almost all P*, which will prove theorem 1.

Clearly, we may assume that $C(K)$ is infinite; let $P_1, ..., P_i, ...$ be its elements.

Lemma 2.

1) Let $f \in H$, $f \neq 0$. Then $\phi_{P_i}(f) \neq 0$ if i is sufficiently large.

2) Let $g \in H$ be another element. Then

$$\frac{|\phi_{P_i}(g)|}{|\phi_{P_i}(f)|} \to \frac{|g|}{|f|}, \quad \text{as} \quad i \to \infty.$$

Let us use the hypothesis that $\mathrm{NS}(C)$ has rank 1. The image of $c_C \in \mathrm{Pic}(C)$ is a basis of $\mathbf{Q} \otimes \mathrm{NS}(C) \cong \mathbf{Q}$. We may thus write in $\mathbf{Q} \otimes \mathrm{Pic}(C)$

(1) $\quad f^* c_A = \lambda(f) c_C + \quad$ class in $\quad \mathbf{Q} \otimes \mathrm{Pic}^\circ(C), \quad$ where $\quad \lambda(f) \in \mathbf{Q}$.

In particular, if $d = \deg(c_C)$ is the degree of the projective embedding of C, we have

(2)
$$\lambda(f) = \frac{1}{d}\deg(f^* c_A) = \frac{1}{d}|f|^2.$$

For $P \in C(K)$, let us compare $h(P)$, the natural height given by the chosen embedding $C \to \mathbf{P}_N$, and $|f(P)|^2$. We have

$$|f(P)|^2 = \tilde{h}_{c_A}(f(P)) = h_{f^* c_A}(P) + O_f(1).$$

The height on C associated to an element of $\mathbf{Q} \otimes \mathrm{Pic}^\circ(C)$ is $o(h(P))$ (cf. §§2.11, 3.10); hence from (1) and (2) we obtain

$$|f(P)|^2 = \frac{1}{d}|f|^2 h(P) + O_f(1) + o(h(P)).$$

This then gives:

For each fixed $f \in H$, we have

$$\frac{|f(P_i)|^2}{h(P_i)} \to \frac{1}{d}|f|^2, \quad \text{as} \quad i \to \infty.$$

Since $\phi_P(f) = f(P)$, we obtain then the two assertions of lemma 2:

1) $\phi_{P_i}(f) = f(P_i) \neq 0$ if i is sufficiently large, because $\tilde{h}(f(P_i)) \neq 0$ for large i.
2) As $i \to \infty$, we have

$$\frac{|\phi_{P_i}(g)|^2}{|\phi_{P_i}(f)|^2} = \frac{\tilde{h}(g(P_i))/h(P_i)}{\tilde{h}(f(P_i))/h(P_i)} \to \frac{\lambda(g)}{\lambda(f)} = \frac{|g|^2}{|f|^2}.$$

It remains to show that ϕ_{P_i} is injective for large i. This follows from the following lemma.

Lemma 3. *Let W_1,W_2 be two Euclidean spaces, Λ a lattice in W_1, and ϕ_i a sequence of linear maps $W_1 \to W_2$ such that for all $f, g \in \Lambda$ with $f \neq 0$*

$$\frac{|\phi_i(g)|}{|\phi_i(f)|} \to \frac{|g|}{|f|}, \quad \text{as} \quad i \to \infty.$$

Then ϕ_i is injective for all large i.

Proof.

Let $e \in \Lambda$, $e \neq 0$. By the hypothesis, we have $\phi_i(e) \neq 0$ for all large i. Replacing ϕ_i by a non-zero multiple, if necessary, we may assume that $|\phi_i(e)| = |e|$. Taking $f = e$, we have $|\phi_i(g)| \to |g|$ for all $g \in \Lambda$. Comparing this result for g,h, and $g + h$, we deduce that for all g,$h \in \Lambda$, then

$$\phi_i(g).\phi_i(h) \to g.h, \quad \text{as} \quad i \to \infty.$$

Let $e_1, ..., e_\lambda$ be a basis of the lattice Λ. We have as $i \to \infty$

$$\det(\phi_i(e_\lambda).\phi_i(e_\mu))_{\lambda,\mu} \to \det(e_\lambda.e_\mu)_{\lambda,\mu} \neq 0,$$

thus $\det(\phi_i(e_\lambda).\phi_i(e_\mu))_{\lambda,\mu} \neq 0$ for large i. Therefore the $\phi_i(e_j), 1 \leq j \leq \lambda$, are linearly independent for i large, and so ϕ_i is injective for large i.

In the next two sections, we give two applications one to quartic curves (due to Demjanenko [D]), the other to modular curves $X_0(p^n)$ (due to Manin [Ma1]).

5.3. First application: Fermat quartics, (cf. [D]).

Let a, b, c be non-zero elements of the number field K, and C the plane quartic

$$ax^4 + by^4 + cz^4 = 0.$$

The genus of C is 3. Its Jacobian $J(C)$ is isogenous to a product of the three elliptic curves $E_\alpha \times E_\beta \times E_\gamma$ defined by the equations

$$v^2 = u^3 - \alpha u, v^2 = u^3 - \beta u, v^2 = u^3 - \gamma u,$$

with

$$\alpha = 4bc/a^2, \beta = 4ac/b^2, \gamma = 4ab/c^2.$$

[This can be seen as follows. Let C_β be the curve of genus 1 defined by

$$ax^4 + by^2z^2 + cz^4 = 0.$$

One has a natural map $(X, Y, Z) \mapsto (XZ, Y^2, Z^2)$ from C to C_β, hence a morphism $\phi_\beta : J(C) \to \mathrm{Jac}(C_\beta)$. By using the formulae in Weil [W4], one checks that $\mathrm{Jac}(C_\beta) \cong E_\beta$. One defines similarly maps $\phi_\alpha : J(C) \to E_\alpha$ and $\phi_\gamma : J(C) \to E_\gamma$ and one checks that they give an isogeny $J(C) \to E_\alpha \times E_\beta \times E_\gamma$, with kernel of type $(2, 2, 2)$.]

Assume now that $a = b = 1$, so that $\alpha = \beta = 4c$. We have rank $\mathrm{Hom}(\mathrm{Jac}(C), E_\alpha) \geq 2$; hence the Manin-Demjanenko theorem applies with $m = 2$ (§5.2, Theorem 2) and gives:

Theorem. *Let $c \in K^*$ be such that the elliptic curve*

$$v^2 = u^3 - 4cu,$$

has K-rank ≤ 1. Then the quartic

$$x^4 + y^4 + cz^4 = 0,$$

has finitely many rational points over K (which can be found effectively).

Remark.

I have not been able to find any concrete example, over $K = \mathbf{Q}$, where the theorem above applies in a *non-trivial* way. [The finiteness of $C(K)$ is trivial if $C(K_v)$ is empty for some completion K_v of K, or if one of the two elliptic curves E_α, E_γ, with $\alpha = 4c, \gamma = 4/c^2$ has rank 0.] In all cases I have tried, either $C(K_v)$ is empty for some v, or rank E_α is even.

[Note that even with the help of Faltings' theorem one still does not know how to find the rational points on a quartic such as $x^4 + y^4 = 17z^4$; are there others than the eight obvious ones $(1, 2, 1), \ldots$?]

Exercise. Let c be a positive integer not divisible by the 4'th power of a prime. Show that there are p-adic points on the curve

$$x^4 + y^4 = cz^4$$

for all p if and only if

a) $c \equiv 1$ or $2 \pmod{16}$,
b) every odd prime factor of c is $\equiv 1 \pmod 8$,
c) $c \not\equiv 3$ or $4 \pmod 5$,
d) $c \not\equiv 7,8$ or $11 \pmod{13}$,
e) $c \not\equiv \pm 6, \pm 7 \pmod{17}$,
f) $c \not\equiv 4, 5, 6, 9, 13, 22, 28 \pmod{29}$.

5.4. Second application: modular curves $X_0(p^n)$, (Manin [Ma1]).

Elliptic curves E equipped with a cyclic subgroup of order N correspond to points on an affine curve $Y_0(N)$ defined over $\mathbf{Q}$, whose function field is $\mathbf{Q}(j(z), j(Nz))$. It can be described as the normalisation of the plane curve given by the equation relating $j(z)$ and $j(Nz)$; here j is the standard modular function.

Let K be a field of characteristic zero. A K-rational point of $Y_0(N)$ corresponds to a pair (E, C) where E is an elliptic curve defined over K, and C is a cyclic group of order N of points of E over the algebraic closure of K, which is stable under the action of $\mathrm{Gal}(\bar{K}/K)$.

More precisely, there is a function $(E, C) \mapsto j(E, C)$ where $j(E, C)$ is a point of $Y_0(N)$; all points of $Y_0(N)$ are obtained thus, and two pairs give the same point: $j(E, C) = j(E', C')$ if (E, C) and (E', C') are isomorphic over $\bar{K}$.

(When $N = 1$, $C = 0$, $j(E, 0)$ is just $j(E)$.)

The curve $Y_0(N)$ is affine, and $X_0(N)$ is its projective completion. Over $\mathbf{C}$, $Y_0(N)(\mathbf{C})$ may be identified with the quotient of the Poincaré upper half plane by

$$\Gamma_0(N) = \{\begin{pmatrix} a & b \\ c & d \end{pmatrix} \in SL_2(\mathbf{Z}); c \equiv 0 \bmod N\};$$

the curve $X_0(N)$ is obtained by adding the "cusps". The genus of this curve tends to infinity with N.

Theorem 1.(Manin). *Let p be a prime number and K a number field. There is an integer $n = n(p, K)$ such that the curve $X_0(p^m)$ has only a finite number of K-rational points for $m \geq n$.*

It is enough to prove that $X_0(p^n)$ has only finitely many K-rational points for some n. Indeed, if $m \geq n$, there is an obvious map

$$X_0(p^m) \to X_0(p^n)$$

which, from the analytic viewpoint, comes from $\Gamma_0(p^m)$ being a subgroup of $\Gamma_0(p^n)$. This map may be described another way: to a pair (E, C_m), where C_m is a cyclic group of order p^m, is associated (E, C_n), where C_n is the cyclic subgroup of C_m of order p^n. Thus if $X_0(p^n)$ has only a finite number of rational points, the same holds for $X_0(p^m)$ for $m \geq n$.

More generally, for $0 \leq i \leq m - n$ there is an arrow $f_i : X_0(p^m) \to X_0(p^n)$ given by

$$(E, C_m) \mapsto (E/C_i, C_{n+i}/C_i).$$

From the analytic point of view, $f_i(z) = p^i z$.

Let n_p be the least integer n such that the genus of $X_0(p^n)$ is at least 1, and let A be the jacobian of $X_0(p^{n_p})$.

[Values of n_p: $n_2 = 5$, $n_3 = 3$, $n_5 = 3$, $n_7 = 2$, $n_{11} = 1$, $n_{13} = 2$, $n_p = 1$ for $p \geq 17$.]

For each $m \geq n_p$, one then has a collection of morphisms, again denoted by f_i:

$$f_i : X_0(p^m) \to A, \quad 0 \leq i \leq m - n_p.$$

Lemma. *If ω is a non-zero invariant differential form on A, the forms $f_i^*(\omega)$, $(0 \leq i \leq m - n_p)$ are linearly independent over $\mathbf{C}$.*

Proof of the lemma.

Expanding with respect to the variable $q = e^{2i\pi z}$, one has

$$\omega = cq^N dq + ..., \; c \neq 0;$$

as f_i transforms q to q^{p^i}, we have

$$f_i^* \omega = cNq^{Np^i + (p^i - 1)} dq + ...;$$

the exponents are distinct for the different values of i; thus the orders of the zero of the $f_i^*\omega$ are distinct, whence the lemma.

From the lemma, the f_i are linearly independent over $\mathbf{Z}$. Let ρ_K be the rank of $A(K)$. If $n \geq n_p + \rho_K$, then the number of f_i is $m - n_p + 1 > \rho_K$, and the Manin-Demjanenko theorem applies, which proves theorem 1.

Theorem 2. *For given p and K, choose $n = n(p, K)$ as in theorem 1. Then there are only finitely many elliptic curves over K (up to K-isomorphism) which have a K-rational point of order p^n (where $n = n(p, K)$).*

Proof.

Note first that $p^n \geq 4$, since otherwise $X_0(p^n)$ would be a projective line.

Let E be an elliptic curve over K having a K-rational point P of order p^n. Let C be the cyclic subgroup generated by P. From theorem 1, there are only a finite number of pairs (E, C) up to $\bar{K}$-isomorphism. Given two pairs (E, P) and (E', P'), there is at most one isomorphism from (E, P) to (E', P') over $\bar{K}$ (because $p^n \geq 4$). By Galois, such an isomorphism is defined over K. Theorem 2 follows easily.

Corollary. *There is an integer $m(p, K) = m$ such that no elliptic curve over K contains a point of order p^m.*

Proof.

From theorem 2, it suffices to take p^m strictly larger than the orders of the p-torsion points of a finite set of elliptic curves.

Conjecture.("Theorem" of Demjanenko). *Let K be a number field. The order of the torsion group of an elliptic curve over K is bounded by a constant depending only on K (and even only on $[K : \mathbf{Q}]$).*

Mazur [Mz2] has proved this for $K = \mathbf{Q}$ with an explicit list of possible orders:$1, 2, 3, ..., 10, 12, 16$ (those for which the corresponding modular curve has genus zero).

[An even more optimistic conjecture is that if A is an abelian variety over $\mathbf{Q}$, one has

$$|A(\mathbf{Q})_{\mathrm{tors}}| \leq C$$

where C depends only on dim A.]

Application to the Tate module of an elliptic curve.

Let E be an elliptic curve over the number field K, let p be a prime number and

$$T_p(E) = \varprojlim E_{p^n}(\bar{K}).$$

the p-adic Tate module of E. Put

$$V_p(E) = \mathbf{Q}_p \otimes T_p(E).$$

The group $\mathrm{Gal}(\bar{K}/K)$ acts on $V_p(E)$.

Theorem. *If the action of* $\mathrm{Gal}(\bar{K}/K)$ *on* V_p *is reducible to triangular form:* $\begin{pmatrix} * & * \\ 0 & * \end{pmatrix}$, *then* E *has complex multiplication.*

[If $\mathbf{Q}(\sqrt{-d})$ is the field of complex multiplication, one can show that p splits in $\mathbf{Q}(\sqrt{-d})$ and that the representation is a direct sum of two representations of degree 1: $\begin{pmatrix} * & 0 \\ 0 & * \end{pmatrix}$, (see [Se3, Ch. IV]).]

Proof.

Assume $V_p(E)$ has a 1-dimensional subspace stable under $\mathrm{Gal}(\bar{K}/K)$. Intersecting this space with $T_p(E)$, gives a $\mathbf{Z}_p$-subspace Γ of $T_p(E)$, of rank 1, which is a direct factor as a $\mathbf{Z}_p$-module and is stable under $\mathrm{Gal}(\bar{K}/K)$. The image C_n of Γ in $T_p(E)/p^nT_p(E) = E_{p^n}(\bar{K})$ is cyclic of order p^n, and stable under $\mathrm{Gal}(\bar{K}/K)$. Let $P_n \in X_0(p^n)$ be the point of $X_0(p^n)$ corresponding to the pair (E, C_n).

From theorem 1, if m is large then $X_0(p^m)$ has only finitely many K-rational points. The points corresponding to $(E/C_i, C_{m+i}/C_i)$ are therefore finite in number; in particular, there is $i < j$ such that $E' = E/C_i$ is isomorphic to $E/C_j = E'/C'_{j-i}$ (where C'_{j-i} is a cyclic subgroup of order $j-i$ of E'). We consider the projection $E' \to E'/C'_{j-i}$:

$$E' \to E'/C'_{j-i} = E/C_j \cong E/C_i = E'.$$

We then have an endomorphism of E' whose kernel is the cyclic subgroup C'_{j-i}. Hence E' has complex multiplication.

For another proof of this theorem, see [Se3, Ch.IV] as well as §8.5. [This is now a special case of Faltings' theorems.]

Appendix. Belyi's theorem.

We have just seen that some special modular curves have few rational points. A theorem of Belyi shows that general modular curves can be almost anything. More precisely:

Theorem. (Belyi, [Be]). *Let X be a non-singular projective curve over $\mathbf{C}$. The following properties are equivalent.*

1) *X is definable over $\bar{\mathbf{Q}}$ (that is to say, X arises by extension of scalars from an algebraic curve over a number field).*
2) *There is a subgroup Γ of $SL_2(\mathbf{Z})$ of finite index such that X is isomorphic to the quotient by Γ of the Poincaré upper half-plane H compactified by adding the cusps.*
3) *There is a finite covering $\pi : X \to \mathbf{P}_1$ unramified outside $\{0, 1, \infty\}$.*

[The proof will also show that if X is defined over $\bar{\mathbf{Q}}$, and S is any finite subset of $X(\bar{\mathbf{Q}})$, the isomorphism

$$X \cong H/\Gamma \cup \{\text{cusps}\}$$

of 2) can be chosen such that S be contained in $\{\text{cusps}\}$.]

Proof.

3) $\Rightarrow$ 1). This is a general fact: every covering of a $\bar{\mathbf{Q}}$-variety can be defined over $\bar{\mathbf{Q}}$ (see e.g. Grothendieck [SGA1]).

2) $\Rightarrow$ 3). As $\Gamma \subset SL_2(\mathbf{Z})$, we have a covering $H/\Gamma \to H/SL_2(\mathbf{Z}) = \mathbf{P}_1 - \{\infty\}$; it is known that this is unramified outside $0, 1728$, and ∞; changing coordinates gives a covering which is unramified outside $0, 1$ and ∞.

3) $\Rightarrow$ 2). Let $\Gamma_1 = SL_2(\mathbf{Z})$ and $\Gamma_2 = (\Gamma_1, \Gamma_1)$, the commutator subgroup of Γ_1. It is known that Γ_2 has index 12 in Γ_1, that it is a free non-abelian group of rank 2, and that

$$H/\Gamma_2 \cong \mathbf{P}_1 - \{a, b, c\},$$

where a, b, c are three distinct points of $\mathbf{P}_1$ (which we may choose to be $0, 1, \infty$). Hence Γ_2 is $\pi_1(\mathbf{P}_1 - \{a, b, c\})$, and every finite (connected, non-empty) covering of $\mathbf{P}_1 - \{a, b, c\}$ is of the form H/Γ, with $\Gamma \subset \Gamma_2$ and $[\Gamma_2 : \Gamma] < \infty$, hence $[\Gamma_1 : \Gamma] < \infty$.

1) $\Rightarrow$ 3). There are two steps.

a) Every curve is a ramified covering of a projective line. Indeed, if C is a curve, f any non-constant rational function on C, one takes $f : C \to \mathbf{P}_1$.

We select f defined over $K \subset \bar{\mathbf{Q}}$. Let $S \subset \mathbf{P}_1(\bar{\mathbf{Q}})$ be the finite set of ramification points of f ("critical values"), i.e. the set of images of the points of C where the differential is zero.

Theorem B. *If $S \subset \mathbf{P}_1(\bar{\mathbf{Q}})$ is a finite set, there is a covering $\phi : \mathbf{P}_1 \to \mathbf{P}_1$ defined over $\bar{\mathbf{Q}}$, and ramified only over $0, 1, \infty$, such that*

$$\phi(S) \subset \{0, 1, \infty\}.$$

This theorem gives 1) $\Rightarrow$ 3), by taking $\pi = \phi \circ f$.

b) *Proof of theorem B.*

If $S \subset \mathbf{P}_1(\bar{\mathbf{Q}})$ is a finite set, and $\phi : \mathbf{P}_1 \to \mathbf{P}_1$ is a non-constant morphism, let S_ϕ denote the union of $\phi(S)$ and the set of critical values of ϕ. We begin with,

Special case: $S \subset \mathbf{P}_1(\mathbf{Q})$. Let $n = |S|$. If $n \leq 3$, there is an automorphism ϕ of $\mathbf{P}_1$, defined over $\bar{\mathbf{Q}}$, such that $\phi(S) \subset \{0, 1, \infty\}$. Suppose now that $n \geq 4$. We shall construct $\phi : \mathbf{P}_1 \to \mathbf{P}_1$, defined over $\mathbf{Q}$ such that $S_\phi \subset \mathbf{P}_1(\mathbf{Q})$ and $|S_\phi| \leq n-1$; by induction on n, this will prove theorem B in this case.

Construction of ϕ. We may suppose that $0, 1, \infty$ and α are in S, α being a rational number distinct from $0, 1, \infty$. We seek ϕ in the form

$$\phi(z) = z^A (z-1)^B,$$

where A and B are rational integers, $A \neq 0$, $B \neq 0$, $A + B \neq 0$. We shall have then $\phi(0), \phi(\infty), \phi(1) \in \{0, \infty\}$. The ramification is given by

$$\frac{d\phi}{\phi} = A\frac{dz}{z} + B\frac{dz}{(z-1)} = 0,$$

i.e. $z = A/(A+B)$. We now choose A and B in $\mathbf{Z}$ such that $\alpha = A/(A+B)$. Then

$$S_\phi \subset \{0, \infty, \phi(\alpha)\} \cup \phi(S'),$$

where $S' = S - \{0, 1, \infty, \alpha\}$, hence $\phi(S')$ has at most $n - 4$ elements.

General case.

We shall deal with the general case by a double induction on the maximum degree over $\mathbf{Q}$ of the elements of S and the number of such elements with this degree.

Let $\alpha \in S$ be a non-rational point, and let $\alpha^n + a_1\alpha^{n-1} + \ldots + a_n = 0$ be its minimal equation over $\mathbf{Q}$. We suppose that n is maximal for the points of S and that the number of such α with this value of n is p. Take

$$\phi(z) = z^n + a_1 z^{n-1} + \ldots + a_n.$$

Then ϕ is defined over $\mathbf{Q}$, and

$$S_\phi \subset \phi(S) \cup \{\infty\} \cup \phi(S')$$

where S' consists of $z \in \bar{\mathbf{Q}}$ such that $\phi'(z) = 0$, hence of elements of degree $\leq n-1$. As the points of $\phi(S)$ have degree $\leq n$ and the number of those of degree exactly n is $\leq p-1$, the induction process works and the theorem follows.

Remark. Manin and Drinfeld ([Dr], [Ma3]) have proved that for a congruence subgroup of $SL_2(\mathbf{Z})$, the cusps on the corresponding curve generate a finite group in the jacobian. Belyi's theorem shows that this becomes false for arbitrary subgroups of finite index.

5.5. The generalised Mordell conjecture.

We consider a group extension of an abelian variety by a group of multiplicative type (a linear torus; the most interesting case is that of an abelian variety). In other words, G is a connected commutative algebraic group not containing $\mathbf{G}_a$. Let X be a subvariety of G and Γ a subgroup of $G(\mathbf{C})$. We suppose that $\mathrm{rank}_{\mathbf{Q}}\Gamma$ is finite.

Conjecture. ([L4]). *There is in X a finite number of translates X_i of subgroups G_i of G such that*

$$X \cap \Gamma = \bigcup_i X_i \cap \Gamma.$$

Consequence. *If X contains no translate of a subgroup of dimension ≥ 1, then $X \cap \Gamma$ is finite.*

This hypothesis is satisfied for a curve X of genus ≥ 2 embedded in its Jacobian.

That Γ has finite rank over $\mathbf{Q}$ does not constrain its torsion. In particular, Γ can contain the full torsion group of G. More generally, if $\gamma_1, ..., \gamma_n$ are points of G, we may take Γ to be all the $g \in G$ for which there is $m \in \mathbf{Z}$, $m \neq 0$, with mg in the subgroup generated by $\gamma_1, ..., \gamma_n$. For $n = 0$, $\Gamma = G_{\mathrm{tors}}$. A consequence is then the following conjecture [now proved by Raynaud [R3]].

Conjecture. (Manin, Mumford). *A curve of genus ≥ 2 embedded in any abelian variety contains only finitely many torsion points.*

[Lang's conjecture is now (1988) known in the following three special cases:

(1) G is a linear torus (Laurent, [Lau]);
(2) G is an abelian variety, and $\dim X = 1$ (Raynaud [R2, Th. 0.3.2], combined with Faltings [F]);

(3) G is an abelian variety and Γ its torsion subgroup (Raynaud, [R3]; for a different proof, based on Galois action, see Hindry [Hn1]).
In particular, the Manin-Mumford conjecture is proved (Raynaud [R1], [R2]).]

5.6. Mumford's theorem [Mu1] **; preliminaries.**

We begin by recalling standard results on divisor classes on a curve C of genus ≥ 1, and on its jacobian J.

Let C be a curve of genus $g \geq 1$ over an algebraically closed field. Let a be a divisor class of C of degree 1. We send C to its jacobian J by the map

$$\phi = \phi_a : C \to J, \quad P \mapsto \mathrm{Cl}(P) - a.$$

Changing the class a alters ϕ by a translation.

If J_d is the principal homogeneous space representing the divisor classes of degree d, there is a canonical map $C \to J_1$, and the identification of J_1 and J corresponds to a choice of a.

The map

$$C \times \ldots \times C = C^g \to J$$

is surjective of degree $g!$ and gives a birational map

$$\mathrm{Sym}^g C \to J.$$

The theta divisor is defined as the image of C^{g-1}:

$$\Theta = \phi(C^{g-1}) = \{\sum_{i=1}^{g-1} \phi(P_i); P_i \in C\}.$$

This is a divisor of J, depending only on a (it is defined intrinsically in J_{g-1}).

The divisor $\Theta' = [-1]^*\Theta$ is given by

$$\Theta' = T_{a'}\Theta, \qquad a' = -K + (2g-2)a. \tag{1}$$

where K is the canonical class. [If one regards Θ as inside J_{g-1}, then $[-1]^*\Theta$ is just $\Theta - K$.]

The proof of (1) uses the Riemann-Roch theorem, cf. [W3].

It is convenient to choose a such that $(2g-2)a \sim K$. Since the canonical class has degree $2g-2$, and an abelian variety is divisible, this is possible at the cost of a finite extension of the base field. It is unique up to the addition of a torsion class, which is unimportant for the associated heights. With this choice of divisor a, Θ is symmetric and the associated normalised height is a quadratic form. Embedding C in its jacobian by ϕ_a then gives a canonical height on C: *on every curve of genus* ≥ 2, *there is one canonical height.*

We consider Θ as an element of $\mathrm{Pic}(J)$. For $c \in J = \mathrm{Pic}^\circ(C)$, let $\phi_{a-c} : x \mapsto \phi_a(x) + c$. In $\mathrm{Pic}(C)$, we have

$$\phi_{a-c}^*(\Theta') = -c + ga.$$

An additivity argument shows that it suffices to prove this for a generic c. We look for the x such that $\phi_a(x) + c \in \Theta'$, that is to say

$$\phi_a(x) + c = (x) - a + c \sim -\sum_{i=1}^{g-1}(P_i) + (g-1)a.$$

Or again

$$(x) + \sum_{i=1}^{g-1}(P_i) \sim -c + ga.$$

As a is of degree 1, and c is generic of degree 0, the divisor $-c+ga$ is generic of degree g; there is then a unique way of writing it as $\sum_{i=1}^{g}(Q_i)$. In order that $\phi_a(x) + c \in \Theta'$ it is necessary and sufficient that x be one of the Q_i's, thus there are g possibilities. One checks that the g points of intersection of the curve with Θ have multiplicity 1, thus

$$\phi_{a-c}^*(\Theta') = \sum_{i=1}^{g}(Q_i) \sim -c + ga.$$

In the particular case of the normalisation with $(2g-2)a = K$, and $c = 0$, we obtain

$$\phi^*\Theta = ga. \tag{2}$$

This shows that Θ intersects C in a divisor of degree g and that the canonical height $\tilde{h}_\Theta$ on J induces on C a height equal to $\frac{g}{2g-2}h_K + O(1)$.

[Recall from §3.4 the following. If A is an abelian variety and X a divisor, the image of X by the automorphism $X \mapsto X + a$ of A is written $T_a X$. We define a homomorphism $\phi_X : A \to A'$ by sending a to the class of $T_a X - X$. Hence the divisor Θ defines a homomorphism ϕ_Θ of J to its dual J'. It is known that the map ϕ_Θ is an isomorphism which identifies J and J'. This is equivalent to the self-intersection of Θ with itself g times being $g!$.]

Let us determine explicitly the Poincaré class $P \in \mathrm{Pic}(J \times J)$ (§3.4). Consider the homomorphisms p_1, p_2 and s from $J \times J$ to J defined by

$$p_1(x, y) = x, \quad p_2(x, y) = y, \quad s(x, y) = x + y.$$

We have the formula

$$P = p_1^*\Theta + p_2^*\Theta - s^*\Theta. \tag{3}$$

This shows in particular that P is symmetric.

We check formula (3) by using the characterisation of the Poincaré divisor given in §3.4.

Put $P_1 = p_1^*\Theta + p_2^*\Theta - s^*\Theta$.

As $x \mapsto T_x X - X$ is a homomorphism in x, we obtain

$$P_1 | J \times \{x\} \sim \Theta + 0 - T_{-x}\Theta \sim T_x\Theta - \Theta.$$

That is to say, $P_1 | J \times \{x\} = x \in \mathrm{Pic}(J)$, since J is identified with its dual via ϕ_Θ. As P_1 induces 0 on $\{c\} \times J$, this implies $P = P_1$.

The morphism $\phi \times \phi$ sends $C \times C$ to $J \times J$. The inverse image of P is given by the formula

$$(\phi \times \phi)^* P \sim \Delta - C \times a - a \times C, \tag{4}$$

where Δ is the diagonal, and where we write (by abuse of notation)

$$C \times a = p_2^* a, \quad a \times C = p_1^* a.$$

Proof of (4). First, if X and Y are non-singular projective irreducible connected varieties, and if $c \in \mathrm{Pic}(X \times Y)$ induces zero on all the verticals and on one horizontal, then $c = 0$ (the analogue of - but distinctly simpler than - the theorem of the cube; see [Mu2, p.54]).

Now, by symmetry it suffices to restrict (4) to a vertical. Let $x_0 \in C$, and let $f: C \to J \times J$:

$$f: C \to C \times C \to J \times J$$

$$x \mapsto (x, x_0) \mapsto (\phi(x), \phi(x_0)).$$

As $P = p_1^*\Theta + p_2^*\Theta - s^*\Theta$, we have

$$f^*P = \phi^*\Theta + 0 - \phi^*_{a-\phi(x_0)}\Theta.$$

But $\phi^*\Theta \sim ga$, (from (2)), therefore

$$f^*P \sim ga + 0 - (ga - \phi_a(x_0)) \sim (x_0) - a,$$

as $\phi_a(x_0) \sim (x_0) - a$.

The right hand side of (4) gives $(x_0) - a$ on $C \times \{x_0\}$. Whence the formula (4).

5.7. Application to heights: Mumford's inequality.

Let K be a field equipped with a product formula, $\bar{K}$ its algebraic closure. We select a (by extending the base field) such that $(2g-2)a = K$, where K is the canonical class, and we embed C in J via $\phi = \phi_a$. As $\Theta = \Theta'$, $\tilde{h}_\Theta$ is a quadratic form on $J(\bar{K})/\text{torsion}$. On the real vector space $V_{\mathbf{R}} = J(\bar{K}) \otimes \mathbf{R}$, this gives a positive quadratic form which we write $h(x) = |x|^2$. We write $x.y$ for the associated scalar product:

$$x.y = \frac{1}{2}(h(x+y) - h(x) - h(y)).$$

Mumford's main result is the following.

Theorem. *There is a constant $M_1(C)$ depending only on C such that if $x, y \in C(\bar{K})$ with $x \neq y$, then*

$$x.y \leq \frac{1}{2g}(|x|^2 + |y|^2 + M_1).$$

Proof.

On $J \times J$, the Poincaré divisor P defines a normalised height and we have

$$\tilde{h}_P(x, y) = -2x.y$$

from the last theorem of §3.4 with $c = \Theta$, (with our conventions $\phi_\Theta(y) = y$). We use (4) of §5.6 : on $C \times C$, we have

$$\tilde{h}_P(x, y) = h_\Delta(x, y) - h_a(x) - h_a(y) + O(1).$$

But $h_\Delta(x, y) \geq O(1)$ outside Δ (§2.10), and as Θ induces ga on C (from (2) of §5.6), h induces gh_a on C; therefore if $x \neq y$

$$\frac{1}{g}|x|^2 + \frac{1}{g}|y|^2 - 2x.y = h_a(x) + h_a(y) + \tilde{h}_P(x, y)$$

$$\geq h_\Delta(x, y) + O(1)$$

$$\geq O(1).$$

As the statement of the theorem is uninteresting for $g = 1$, *we assume from now on that* $g \geq 2$.

Corollary. *There is a constant* $M_2 = M_2(C)$ *such that if* x *and* y *are points of* $C(\bar{K})$ *with* $M_2 \leq |x| \leq |y| \leq 2|x|$ *and* $x \neq y$ *then the angle* $\theta_{x,y}$ *between the vectors* x,y *in* $V_{\mathbf{R}}$ *is* $\geq \frac{\pi}{6}$.

Proof. We write $x.y = |x|.|y|\cos(\theta)$; as $g \geq 2$ we have

$$\cos(\theta) \leq \frac{1}{4}\left(\frac{|x|}{|y|} + \frac{|y|}{|x|}\right) + \frac{M_1}{|x|.|y|}.$$

Take $M_2 = \sqrt{10 M_1}$. Then

$$\cos\theta \leq \frac{3}{4} + \frac{M_1}{|x|^2} \leq \frac{3}{4} + \frac{1}{10} = \frac{85}{100}.$$

But

$$\cos\frac{\pi}{6} = \frac{\sqrt{3}}{2} = 1.732.../2 = 0.86... > 85/100 \geq \cos\theta,$$

therefore $\theta > \frac{\pi}{6}$.

[Note that if $g \geq 3$ one even obtains $\theta \geq \frac{\pi}{3}$.]

We call a point $x \in C(\bar{K})$ "large" if

$$|x| = h(x)^{1/2} > M_2.$$

The space $V_{\mathbf{R}}$ has infinite dimension. It is easily seen that the points $x \in C(K)$ where K is a number field lie in a finite dimensional subspace of $V_{\mathbf{R}}$.

For an angle θ with $0 < \theta < \pi/2$, denote by $N(n, \theta)$ the maximum number of points $x_1, ..., x_N$ in the unit sphere of the Euclidean space $\mathbf{R}^n$ whose mutual angles are $\geq \theta$. For $\theta = \pi/3$, there is a sphere packing interpretation: $N(n, \pi/3)$ is the maximum number of mutually disjoint unit spheres tangential externally to a given unit sphere ("kissing number").

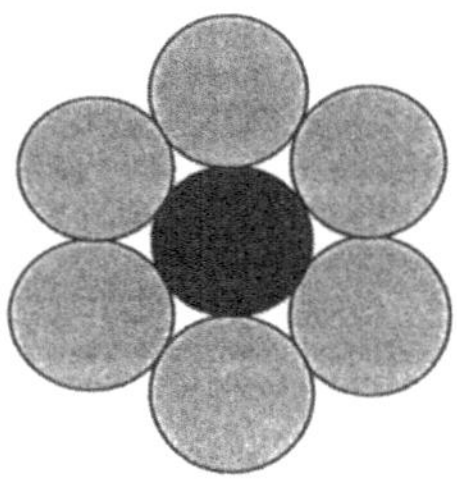

This is an interesting function of n. For small values of n, one has

$$N(1, \pi/3) = 2; \quad N(2, \pi/3) = 6; \quad N(3, \pi/3) = 12, \quad N(4, \pi/3) = 24 \text{ or } 25.$$

Odlyzko and Sloane have shown that $N(8, \pi/3) = 240$, and $N(24, \pi/3) = 196560$. For $n = 16$, it is only known that

$$4320 \leq N(16, \pi/3) \leq 8313.$$

We use the following rough bound.

Theorem. $N(n, \frac{\pi}{6}) \leq 5^n$.

Proof.

Let $x_1, ..., x_N$, with $N = N(n, \frac{\pi}{6})$, be points of the unit sphere of $\mathbf{R}^n$, $n \geq 1$, whose mutual angles are $\geq \frac{\pi}{6}$. We remark that the spheres with centres x_j and of radius $\frac{1}{4}$ do not intersect.

For this, it suffices to see that $\sin \frac{\pi}{12} > \frac{1}{4}$; indeed, we have

$$\sin \frac{\pi}{12} = \frac{1}{2} \frac{\sin \frac{\pi}{6}}{\cos \frac{\pi}{12}} = \frac{1}{4} \frac{1}{\cos \frac{\pi}{12}} > \frac{1}{4}.$$

The unit sphere having a volume v, say, the spheres of radius $\frac{1}{4}$ have volume $(\frac{1}{4})^n v$ and are all contained in a sphere of radius $5/4$, and of volume $(5/4)^N v$. Thus $N.(\frac{1}{4})^n v < (5/4)^n v$ and therefore $N < 5^n$.

Theorem. *Let $\rho_K = \operatorname{rank} J(K)$. Then the number of large points x of $C(K)$ such that $|x| \leq A$ (with $A \geq M_2$) is*

$$\leq 5^{\rho_K}(1 + \frac{\log(A/M_2)}{\log 2}).$$

Indeed, let n be such that $2^n M_2 \leq A < 2^{n+1} M_2$. If x, y are distinct points of $C(K)$, such that $|x|$ and $|y|$ belong to the same interval $(2^j M_2, 2^{j+1} M_2)$, $0 \leq j \leq n$, the corollary above shows that the angle $\theta_{x,y}$ between these points is $\geq \pi/6$; consequently, the number of such points is $\leq 5^{\rho_K}$, by the previous theorem. Since there are $n+1$ intervals, the total number of large points with $|x| \leq A$ is

$$\leq (n+1)5^{\rho_K} \leq 5^{\rho_K}(1 + \frac{\log A/M_2}{\log 2}).$$

Corollary. *The number of points x of $C(K)$ with $h(x) \leq A$ is*

$$\leq 5^{\rho_K} \frac{\log A}{2 \log 2} + \text{const.}, \quad \text{as } A \to \infty.$$

Therefore the number of points of $C(K)$ of (exponential) height $H(x) \leq A$ is $\leq O(\log \log A)$.

Remark. Indeed, the result obtained [although superseded by Faltings for number fields] is optimal for function fields over finite fields.

Let Y be an algebraic curve over $\mathbf{F}_q$, $K = \mathbf{F}_q(Y)$, and C a curve defined over $\mathbf{F}_q$. The elements of $C(K)$ are morphisms defined over $\mathbf{F}_q$ of Y in C; for $f \in C(K)$, the height of f is

$$h(f) = |f|^2 = \lambda \deg f.$$

The constant morphisms (which are finite in number) are the f's of degree zero. If f is a non-constant point, $|f| \neq 0$, the composites with the Frobenius endomorphisms π_q^n are:

$$f_n = \pi_q^n \circ f = f \circ \pi_q^n.$$

Then $\deg f_n = q^n \deg f$, and the number of points f_n of height $\leq A$ is $\asymp \log A$.

6. LOCAL CALCULATION OF NORMALISED HEIGHTS

Néron [N4] has shown that the canonical height over a number field can be broken as a sum of local heights. This leads to the question of calculating these local heights both for the prime numbers and for the infinite primes.

Tate [T2] has made this theory explicit for the case of elliptic curves in two letters to Serre; the first letter has been reproduced by Lang [L5] and taken up again by Zimmer [Z].

We only sketch the results; for complete proofs the reader should look at [N4] or [L5].

6.1. Bounded sets.

Let K be a complete normed field (K will be $\mathbf{R}$, or $\mathbf{C}$, or a complete field for a real valuation).

We write

$$v(x) = -\log |x|_v.$$

Let V be an algebraic variety over K. Besides the Zariski topology on $V(K)$, we have a topology induced by that of K, which is called the K-topology.

We now define the notion of a *bounded* subset of $V(K)$, first in the affine case, then in the general case.

(1) Assume that V is affine. The affine algebra of V is of finite type over K; we write this as

$$K[V] = K[f_1, ..., f_m].$$

A subset B of $V(K)$ is called *bounded* if $f_i(B)$ is bounded in K for all i. This condition is independent of the choice of generators $f_1, ..., f_m$ of $K[V]$.

(2) General case. Cover V by a finite number of open affines: $V = \bigcup_i U_i$. A subset B of $V(K)$ is called bounded if there are $B_i \subset U_i(K)$ which are bounded and $B = \bigcup_i B_i$.

Definitions (1) and (2) are readily seen to be compatible.

When K is locally compact, bounded is equivalent to relatively compact for the K-topology.

Projective space is bounded (and so also is any projective variety). To see this take homogeneous coordinates $(x_0, ..., x_n)$, and define

$$W_i = \{x; |x_i| \geq |x_j| \quad \text{for all} \quad j\}$$

The W_i cover $\mathbf{P}_{n/K}$ and W_i is bounded in the open affine given by $x_i \neq 0$: in affine coordinates x_j/x_i, $(j \neq i, 0 \leq j \leq n)$, W_i is the unit cube.

For example, the projective plane is a union of 3 bounded subsets.

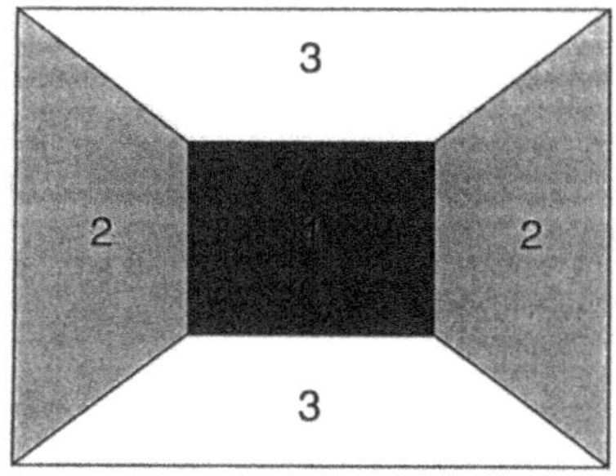

Here are some elementary properties of bounded sets:-

(1) Let $f : V \to W$ be a morphism of K-varieties. If $B \subset V(K)$ is bounded then $f(B)$ is bounded in $W(K)$.

(2) Let V be a closed subvariety of W and $B \subset V(K)$. Then B is bounded in $V(K) \Leftrightarrow B$ is bounded in $W(K)$.

(3) Let $f : V \to W$ be a proper morphism of quasi-projective varieties. If $B \subset W(K)$ is bounded then $f^{-1}(B)$ is bounded in $V(K)$.

(4) Let V be an affine variety, W a closed subvariety of V and $V' = V - W$. Let ϕ_α be a finite set of generators of the ideal defining W. Then $B \subset V'(K)$ is bounded in $V'(K)$ if and only if
a) B is bounded in $V(K)$,
and
b) $\inf_\alpha 1/|\phi_\alpha|$ is bounded on B.

(5) Let V be a closed subvariety of $\mathbf{P}_n$ and W a closed subvariety of V defined by homogeneous equations $\phi_\alpha(x_0, ..., x_n) = 0$ of degree d_α and $V' = V - W$. If $x \in V(K)$, $\phi_\alpha \neq 0$, put

$$b_\alpha(x) = \sup_i |x_i^{d_\alpha}/\phi_\alpha(x_0, ..., x_n)|;$$

If $x \in V'(K)$ put $b(x) = \inf_\alpha b_\alpha(x)$. A subset $B \subset V'(K)$ is bounded if and only if the function b is bounded on B.

Exercise. Assume that K is a complete field with respect to a discrete valuation; let O_K be its ring of integers and π a uniformizing element.

Let V be a closed subvariety of affine n-space $\mathbf{A}_n$ and $\bar{V}$ its Zariski closure in $\mathbf{P}_n$. Suppose $\bar{V}(K) = V(K)$, i.e. $\bar{V}$ has no K-point at infinity. Prove that $V(K)$ is bounded in $\mathbf{A}_n$. [When K is locally compact, this follows from a compactness argument. In the general case, use a theorem of Greenberg [Gr] according to which an O_K-scheme of finite type has an O_K-point if and only if it has an $O_K/\pi^n O_K$-point for every $n \geq 1$.]

6.2. Local heights.

Let V, K be as above.

Definition. We say that a real valued function f on $V(K)$ is *strongly continuous* if it is
1) continuous,
2) bounded above and below on all bounded subsets of $V(K)$.

When K is a locally compact field, the first condition implies the second, thus strongly continuous is equivalent to continuous (this is false if K is not locally compact).

If a function on $V(K)$ induces strongly continuous functions on an affine open covering, then it is strongly continuous.

Assume now that V is quasi-projective and D is a Cartier divisor of V with support Supp D (if V is non-singular, $D = \sum n_W.W$ where W has codimension 1, cf. §2.7).

It will be convenient to assume that $V(K) - \text{Supp } D$ is dense in $V(K)$, which is true if K is algebraically closed (which is a harmless assumption) or if V is smooth (for example an abelian variety).

The local height is a real valued function

$$\lambda : V(K) - \text{Supp } D \to \mathbf{R},$$

which satisfies the following condition. If U is an open affine of V on which D is a principal divisor (ϕ), where ϕ is a rational function on U, then the function

$$x \mapsto \lambda(x) - v(\phi(x)),$$

defined on $U(K) - \text{Supp } D$, extends to a strongly continuous function (in the above sense) on $U(K)$.

If D is a positive divisor with local equation ϕ, then $v(\phi) = -\log|\phi|$ tends to ∞ near the divisor, and λ is the sum of $v(\phi)$ and a strongly continuous function.

We say that λ is associated to D. If λ exists, it is unique up to the addition of a strongly continuous function on V (the difference of two such λ's is a strongly continuous function on a covering, thus is strongly continuous on V).

Proposition. *Local heights exist.*

Example. If $V = \mathbf{P}_n$, D is the hyperplane $t = 0$ in $\mathbf{P}_n$, where

$$t = \sum_{i=0}^{n} \alpha_i x_i,$$

(for instance $t = x_0$), then

$$\lambda_t(x) = v(t) - \inf_{0 \leq i \leq n} v(x_i),$$

is a local height associated to D.

Proof of the proposition.

By enlarging K we may assume that it is algebraically closed.

Notice that

a) if D is a principal divisor (ϕ) we may take $\lambda_D = v(\phi)$,

b) if $D = D_1 + D_2$ and $\lambda_{D_1}, \lambda_{D_2}$ are associated with D_1, D_2, then $\lambda_{D_1} + \lambda_{D_2}$ is associated to $D_1 + D_2$.

It is known that the line bundle defined by D is a difference of ample line bundles (see the proof of the theorem in §2.8). Hence using a) and b) one is reduced to the case where $D = f^{-1}(H)$, for $f : V \to \mathbf{P}_n$, H being a hyperplane of $\mathbf{P}_n$. In that case, one may take λ_D to be $\lambda_H \circ f$ where λ_H is a local height attached to H, see above.

Examples.

1) *Trivial examples.*

a) If D is a principal divisor, $D = (\phi)$, take $\lambda = v(\phi(x))$. Even if the variety is projective, λ is not canonical, for ϕ is only defined up to multiplication by a constant.

b) Let $D = D_1 + D_2$. We suppose that λ_{D_1} and λ_{D_2} are associated to D_1 and D_2, respectively; then $\lambda_{D_1} + \lambda_{D_2}$ is defined outside Supp $D_1 \cup$ Supp $D_2 \supset$ Supp D, and extends by continuity outside of Supp D to a function λ attached to D.

2) *Over* $\mathbf{C}$. A Cartier divisor is the same as a line bundle together with a meromorphic section of this bundle (§2.7). Put a hermitian structure on this line bundle (first locally, then use a partition of unity); this gives a measure of the length of sections of this bundle. The absolute value of a section gives a real-valued function on $V(\mathbf{C})$ − Supp D (which one can

extend by 0 at the zeros of the section and which is not defined at the poles). A function λ satisfying the definition is

$$\lambda(x) = -\log|s(x)|.$$

3) *K locally compact.* The above construction can be imitated. We have a line bundle whose structural group is the group K^*; the absolute value $|.| : K^* \to \mathbf{R}^*$ then gives a topological line bundle with $\mathbf{R}^*_+$ as its structural group (in the ultrametric case K^* is totally disconnected and the bundle is trivial). Put a hermitian structure on this bundle, as above, and take the same construction.

4) *Affine varieties.* Let V be an affine variety. A positive divisor D corresponds to an ideal $\mathbf{d}$ of the affine algebra $K[V]$. Let $(Q_1, ..., Q_n)$ be a system of generators of $\mathbf{d}$. We put

$$\lambda(x) = \inf_i v(Q_i(x)),$$

which makes sense except if all the $Q_i(x)$ are zero, that is to say outside of Supp D. To check that λ is associated to D, we must show that if $D = (\phi)$, then $\lambda(x) - v(\phi(x))$ is regular.

But if $D = (\phi)$, we have $Q_i = P_i\phi$, and the ideal generated by the P_i is 1: $\sum_i R_i P_i = 1$; that the function $\inf_i v(P_i)$ is regular is then a simple exercise.

5) *Projective case.* We embed the variety V in $\mathbf{P}_n$ with homogeneous coordinates $(x_0, ..., x_n)$. The positive divisor D is defined by an homogeneous ideal $\mathbf{d}$, generated by a system of homogeneous elements (P_α), with $\deg P_\alpha = d_\alpha$. We then put

$$\lambda(x) = \inf_\alpha \{v(P_\alpha(x)) - d_\alpha \inf_{0 \leq i \leq n} v(x_i)\}.$$

One checks that multiplying the x_i by a constant leaves $\lambda(x)$ invariant and that $\lambda(x)$ is defined outside Supp D.

That λ is associated to the divisor D is seen by examining each open affine $x_i \neq 0$; the change to affine coordinates recovers the formula of the affine case.

6) *Local heights as intersection multiplicities.* Suppose that K has a discrete valuation and let the ring of integers be O_K. Take a scheme $\mathbf{V}$ projective

and smooth over O_K, with absolutely irreducible fibres, and denote by V the variety over K derived from this scheme (i.e. V is the generic fibre of $\mathbf{V}$).

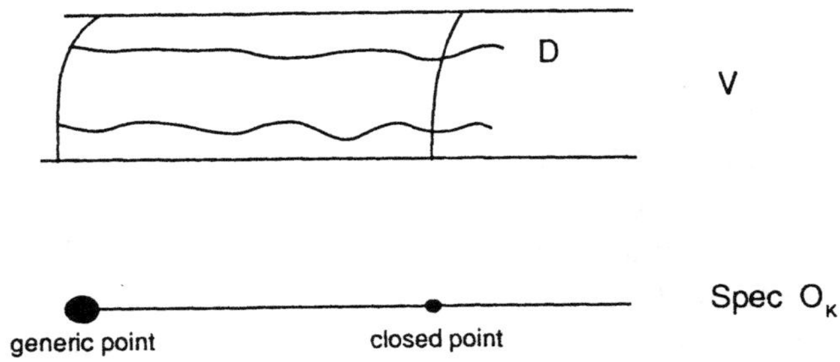

By taking the Zariski closure, a divisor D on V extends to a divisor $\mathbf{D}$ on $\mathbf{V}$ and in the same way a point $x \in V(K)$ extends to an integral point $\mathbf{x} \in \mathbf{V}(O_K)$.

Assume that $x \notin \operatorname{Supp} D$. Define $m(x, D)$ as follows:

a) If $\mathbf{x}$ does not intersect $\mathbf{D}$, then $m(x, D) = 0$.

b) If $\mathbf{x}$ does intersect $\mathbf{D}$ in a point x_0 of the special fibre, then $m(x, D)$ is the *intersection multiplicity* of $\mathbf{x}$ and $\mathbf{D}$ and x_0 (computed on $\mathbf{V}$; note that if V has dimension n, then $\mathbf{V}$ has dimension $n+1$, $\mathbf{D}$ has dimension n, and $\mathbf{x}$ has dimension 1).

[One way to compute $m(x, D)$ in case b) is to write $\mathbf{D}$ as (ϕ) at x_0, for some ϕ, and put $m(x, D) = v(\phi(x))$. Another way is to view $\mathbf{x}$ as a morphism $\operatorname{Spec}(O_K) \to \mathbf{V}$ and take the inverse image of $\mathbf{D}$, which is an integral multiple $m(x_0)$ of x_0; then $m(x, D) = m(x_0)$.]

We thus get a function $x \mapsto m(x, D)$, and one checks that it is a *local height* associated to D.

Exercise. Let V be a variety, $\mathbf{U} = \{U_1, ..., U_n\}$ a finite open affine covering of V, Λ_i the affine ring of U_i, Φ_i a finite subset of Λ_i containing 1 and generating Λ_i as a K-algebra. Assume that K is complete for a non-archimedean absolute value $|.|$.

(1) If $x \in U_i(K)$, define $\epsilon_i(x) = \sup_{\phi \in \Phi_i} |\phi(x)|$. Define $\Psi_i \subset U_i(K)$ as the set of $x \in U_i(K)$ such that $\epsilon_i(x) \leq \epsilon_j(x)$ for all j such that $x \in U_j(K)$. Prove that Ψ_i is open and closed in $X(K)$ (use the properties of bounded sets in §6.1).

(2) If $x \in V(K)$, define $i(x)$ to be the smallest i with $x \in \Psi_i$, and let $\Omega_i = \{x | i(x) = i\}$. Then $V(K)$ is the disjoint union of the Ω_i, which are open and closed in $V(K)$. Show that if $i, j_1, ..., j_k$ are any indices and B is bounded in $U_{j_1}(K) \cap ... \cap U_{j_k}(K)$, with respect to $U_{j_1} \cap ... \cap U_{j_k}$, then $\Omega_i \cap B$ is bounded

with respect to $U_i \cap U_{j_1} \cap ... U_{j_k}$.
(3) Show that the cohomology groups $H^k(\mathbf{U}, \mathbf{S})$ are zero for $k \geq 1$, where $\mathbf{S}$ is the sheaf of strongly continuous functions.

(If $(f_{i_0 \ldots i_k})$ is a k-cocycle, show that it is the coboundary of $g_{i_1 \ldots i_k}$ defined by

$$g_{i_1 \ldots i_k}(x) = f_{i i_1 \ldots i_k}(x), \text{ where } i = i(x).)$$

(4) Use the vanishing of $H^1(\mathbf{U}, \mathbf{S})$ to give another proof of the existence of local heights.

In particular, if D is a positive divisor on V such that $D = (f_i)$ on each U_i $(f_i \in \Lambda_i)$, show that λ_D can be taken to be

$$\lambda_D(x) = v(f_i(x)), \text{ where } i = i(x).$$

6.3. Néron's theorem.

We now consider abelian varieties. Here is a first normalisation result.

Theorem. (Néron [N4]). *To all pairs (A, D), where A is an abelian variety over K and D a divisor on A, there is an associated local height*

$$\lambda_D : A(K) - \text{Supp } D \to \mathbf{R},$$

(defined up to the addition of a constant function), such that:
(1) If $D = (\phi)$, $\lambda_D(x) = v(\phi(x))$ up to a constant.
(2) $\lambda_{D_1} + \lambda_{D_2} = \lambda_{D_1+D_2}$ modulo constants
(3) If $\pi : B \to A$ is a morphism of abelian varieties and if $D_B = \pi^* D$ is defined (i.e. if $\pi(B) \not\subset$ Supp D), then $\lambda_{\pi^* D} = \lambda_D \circ \pi$ up to a constant (we do not assume that π maps 0 to 0).

Proof (abridged).

We may suppose that the class of D in $\text{Pic}(A)$ is symmetric (resp. antisymmetric). Under the map $x \mapsto 2x$ on A, D becomes (cf.§3.2 Corollary)

$$2^* D = rD + (\phi), \quad \text{with } r = 4 \text{ (resp. 2)}$$

where ϕ is a rational function. If μ is a local height associated to D, we have

$$\mu(2x) = r\mu(x) + v(\phi(x)) + \epsilon(x)$$

where $\epsilon(x)$ is strongly continuous on $A(K)$ (the two local heights $\mu(2x)$ and $r\mu(x) + v(\phi(x))$ are associated to $2^* D$ and $rD + (\phi)$, respectively). From a general fact on Banach spaces (cf. §3.1, Exercise) there is a strongly continuous function η such that $\epsilon = \eta(2x) - \left\{ \begin{matrix} 4 \\ 2 \end{matrix} \right\} \eta(x)$. Put $\lambda = \mu - \eta$. This is then a local height associated to D, and

$$\lambda(2x) = r\lambda(x) + v(\phi(x)).$$

(The normalised function to be found had to satisfy this formula.)

One then shows that this is the appropriate local height. The uniqueness up to the addition of a bounded function (this ambiguity arises from the choice of ϕ) results from the following elementary lemma.

Lemma. *If $\nu(2x) = r\nu(x)$ and ν is bounded, then $\nu = 0$.*

We call the function λ_D of the theorem the normalised local height (even so, it is only defined up to an additive constant).

Example 1. Computation over **C** *via theta functions.*

Let A be an abelian variety over **C**. This is a quotient of a **C**-vector space V by a lattice Γ (cf. §3.2). A "theta function on A" is a meromorphic function θ on V which satisfies the identity

$$\theta(x+\gamma) = \theta(x) e^{2\pi i \{L(x,\gamma) + H(\gamma)\}}, \quad \gamma \in \Gamma,$$

where L is **C**-linear in x (further normalisation is possible).

The divisor of θ is invariant under translation by Γ hence may be identified with a divisor of A.

Theorem. *Every divisor of A is a divisor of a theta function, defined up to multiplication by a trivial theta function $e^{Q(x)}$ (where Q is a quadratic polynomial in x).*

(For the proof, see e.g. [Sw, Theorem 18] or [Mu2, Chapter1].)

Calculating $\theta(x + \alpha + \beta)$ in the two obvious different ways gives some relations between the exponents. This type of argument shows that the function $L(x, \gamma)$ extends to a function $L(x, y)$ linear in y, and that there is a **R**-quadratic polynomial R in x such that the function

$$\tilde{\theta}(x) = |\theta(x)| e^{R(x)}$$

satisfies $\tilde{\theta}(x+\gamma) = \tilde{\theta}(x)$. Hence $\tilde{\theta}$ is a positive real valued function on $A(\mathbf{C})$. We then put

$$\lambda(x) = v(\tilde{\theta}(x)).$$

It is defined outside $D = \operatorname{div}(\theta)$, and one checks that it has properties (1),(2),(3) of Néron's theorem with respect to D; hence it is the *normalised local height relative to* D.

Example 2. Good reduction.

Suppose that K has a discrete valuation v and A/K has good reduction at v. Then the closed fibre of the Néron model $\mathbf{A}$ over O_K is again an abelian variety and the intersection multiplicity $\lambda_D(x) = m(x, D)$ (cf.§6.2.(6)) is a normalised local height.

6.4. Relation with global heights.

As in §2.1, suppose K is a field equipped with a product formula with local exponents n_v.

Let D be a divisor on A and $\tilde{h}_D$ the corresponding canonical height.

Theorem. *There is a choice of $\lambda_{D,v}$ where $\lambda_{D,v}$ is a normalised local height associated to D over the local field K_v, such that for all $x \notin$ Supp D*

$$\tilde{h}_D(x) = \sum n_v \lambda_{D,v}(x),$$

where for each x all except finitely many terms of the sum are zero.

We only sketch the proof; for a more complete account see [L6, Chapter11].

The abelian variety has good reduction at almost all places. For such places, extending the divisor so that it has no component on the special fibre gives a canonical $\lambda_{D,v}$. We have $\lambda_{D,v}(x) = 0$ for almost all v. For the finite number of bad reduction places v, we choose any $\lambda_{D,v}$, and we must show that the formula is correct up to the addition of a constant.

Let $x \in A(K)$; select ϕ such that the divisor $D' = D + (\phi)$ does not go through x, and let $\lambda_{D',v}$ be the unique normalised local height such that

$$\lambda_{D',v} = \lambda_{D,v} + v(\phi), \quad \text{outside Supp } D \cup \text{Supp } D'.$$

Put

$$h'(x) = \sum_v n_v \lambda_{D',v}(x);$$

it does not depend on the choice of ϕ. In particular, if $x \notin$ Supp D, one can take $\phi = 1$ hence

$$h'(x) = \sum_v n_v \lambda_{D,v}(x), \quad x \notin \text{Supp } D.$$

By changing one of the $\lambda_{D,v}$, it may be assumed that $h'(0) = 0$.

Then one checks:

a) h' is a global height relative to D (i.e. it is a "h_c" attached to the class c of D, cf. §2.8),

b) $h'(2x) = 4h'(x)$ if D is symmetric (resp. $h'(2x) = 2h'(x)$ if D is antisymmetric).

The theorem follows.

6.5. Elliptic curves.

Let E be an elliptic curve over a local field K and D a divisor on E. To define a local height λ associated to D, by translation and additivity we may reduce to the case where $D = (0)$. The function λ will be continuous except at 0, and bounded outside every neighbourhood of 0. Further, if z is a local coordinate at 0,

$$\lambda(P) - v(z(P)) \to \text{const.}, \quad \text{as } P \to 0.$$

We shall completely normalise λ. There are two ways of doing this, both suggested by Tate.

First normalisation. Select a differential form of the first kind $\omega \neq 0$ on E and choose z such that $\omega = dz+$(terms of higher degree) about 0. We then demand that,

$$\lambda(P) - v(z(P)) \to 0, \quad \text{as } P \to 0.$$

This depends only on ω, not on z. The corresponding local height will be denoted λ_ω.

Second normalisation. Choose ω as above; let Δ be the discriminant of the corresponding Weierstrass equation. Put

$$\lambda(P) = \lambda_\omega(P) + \frac{1}{12}v(\Delta).$$

One checks that λ is independent of the choice of ω (it is equal to λ_ω when ω is such that Δ is a unit).

There are three types of computations to do.

1. Non-archimedean calculations; there are two cases:
   - Curves with good reduction. In this case one has $\lambda(P) = 0$ if the reduction $\tilde{P}$ of P is $\neq \tilde{0}$, and $\lambda(P) = v(z(P))$ if $\tilde{P} = \tilde{0}$, where $z = x/y$ is a local parameter normalised with respect to the differential form.

   - Tate curves, with multiplicative bad reduction.

These curves are dealt with by means of theta functions. In the Tate model, the curve is expressed as $E = \mathbf{G}_m/q^{\mathbf{Z}}$ with $0 < |q| < 1$. The function

$$\theta(t) = (1-t)\prod_{n=1}^{\infty}(1-q^n t)(1-q^n t^{-1})$$

satisfies the identity

$$\theta(qt) = -\frac{1}{t}\theta(t).$$

As in the complex case, one corrects the absolute value of θ by a quadratic polynomial

$$\lambda_\omega(t) = v(\theta(t)) + \frac{\alpha^2 - \alpha}{2} v(q)$$

where $\alpha = v(t)/v(q)$, for the natural choice $\omega = dt/t$. As

$$\Delta = q\prod(1-q^n)^{24},$$

we have $v(\Delta) = v(q)$, and then the absolute λ is

$$\lambda(t) = v(\theta(t)) + \frac{1}{2}(\alpha^2 - \alpha + \frac{1}{6})v(q).$$

Note that $(X^2 - X + \frac{1}{6})$ is the second Bernoulli polynomial $B_2(X)$, which is an interesting feature of this normalisation.

2. Calculation over $\mathbf{C}$.

One possible method is to use a theta function, for example the Weierstrass sigma function (where we use the standard notation)

$$\sigma(z) = z\prod_{w\neq 0}(1-\frac{z}{w})e^{\frac{z}{w}+\frac{1}{2}\frac{z^2}{w^2}}.$$

Then

$$\lambda_\omega(z) = -\log|\sigma(z)| + \frac{1}{2}\mathrm{Re}(z\eta(z)),$$

where η is the $\mathbf{R}$-linear function $\mathbf{C} \to \mathbf{C}$ which transforms the fundamental periods ω_1, ω_2 to η_1, η_2 and ω is the standard differential form dz.

Practical computation by this method is inconvenient. But there is a better one, due to Tate.

With the usual notation:

$$y^2 + a_1xy + a_3y = x^3 + a_2x^2 + a_4x + a_6$$

we take

$$\omega = \frac{dx}{2y + a_1x + a_3}.$$

One has

$$\lambda_\omega(P+Q) + \lambda_\omega(P-Q) = 2\lambda_\omega(P) + 2\lambda_\omega(Q) + v(x(P) - x(Q)),$$

$$\lambda_\omega(2P) = 4\lambda_\omega(P) + v((2y + a_1x + a_3)(P)),$$

and for the canonical λ

$$\lambda(P+Q) + \lambda(P-Q) = 2\lambda(P) + 2\lambda(Q) + v(x(P) - x(Q)) - \frac{1}{6}v(\Delta),$$

$$\lambda(2P) = 4\lambda(P) + v((2y + a_1x + a_3)(P)) - \frac{1}{4}v(\Delta).$$

These formulae follow from an equivalence of the corresponding divisors, for example:

$$2^*(0) = 4.(0) + (2y + a_1x + a_3),$$

shows that the formula for $\lambda_\omega(2P)$ is true up to a constant; checking the behaviour of each side near 0 shows that the constant is zero.

Tate's recipe for numerical calculation is the following. Put

$$Y = 2y + a_1x + a_3$$

so that the equation of the curve becomes

$$Y^2 = 4x^3 + b_2x^2 + 2b_4x + b_6,$$

and define Z by

$$x(2P) = \frac{x^4 - b_4x^2 - 2b_6x - b_8}{4x^3 + b_2x^2 + 2b_4x + b_6} = x^4\frac{Z}{Y^2},$$

that is to say

$$Z = 1 - b_4 x^{-2} - 2b_6 x^{-3} - b_8 x^{-4}.$$

The formula

$$(*) \qquad \lambda(P) = \frac{1}{2}\log|x(P)| + \frac{1}{8}\sum_{n=0}^{\infty}\frac{1}{4^n}\log|Z(2^nP)|.$$

has been proved by Tate (see [T2] and [Z]) in the case when $K = \mathbf{R}$ and all real points on the curve satisfy $x > 0$. In this case Z is a rational function without zeros or poles on the real points, thus its absolute value is bounded above and below. Then $\log|Z(2^nP)|$ is bounded in absolute value and the series converges rapidly.

In the general case, Z has zeros and poles, and 2^nP can approach a singular point. Nevertheless, one conjectures that this formula (*) remains valid [it has been proved by Masser [Mas] in the case of complex multiplication using elliptic analogues of Baker's theorem].

7. SIEGEL'S METHOD

In the next two chapters we study integral points on curves, first with Siegel's (ineffective) method and second with Baker's effective method of linear forms in logarithms.

7.1. Quasi-integral sets.

We begin with a particular case. Let $F(x, y) = 0$ be an irreducible plane curve X over $\mathbf{Q}$. Siegel's theorem says that the integral points $(x, y) \in \mathbf{Z}\times\mathbf{Z}$ on this curve are finite in number, except when the curve has genus zero and has either 1 or 2 points at infinity.

Here are examples of such exceptional curves. The curve $y = 0$ is an affine line having infinitely many integral points $x \in \mathbf{Z}$; it has one point at infinity.

The conic $y^2 - Dx^2 = 1$, where D is a non-square positive integer, has two points at infinity. It is well known that this Pell-Fermat equation has infinitely many integral solutions.

Over an arbitrary number field, another exceptional case arises: $xy = 1$, where the integral points correspond to units in the number field.

To give the general statement, we take a number field K (of finite degree over $\mathbf{Q}$), S will be a finite set of places of K containing the infinite places, and $A = A_S$ denotes the ring of S-integers of K:

$$A_S = \{\alpha \in K; |\alpha|_v \leq 1 \text{ for all } v \notin S\}.$$

Let X be an affine variety over K, $\Lambda_X = \mathrm{H}^0(X, O_X)$ the affine algebra of X, and M a set of rational points of X: $M \subset X(k)$. We shall say that *M is quasi-integral relative to the ring A* if for all $f \in \Lambda_X$ there is $a \in K^*$ such that $f(M) \subset aA$ (that is to say, the 'denominators of $f(x)$, $x \in M$ are bounded').

It suffices to check this condition for a set of generators of the K-algebra Λ_X. In particular, if we select an immersion of the variety $X \subset \mathbf{A}_n$, with some coordinates $z_1, ..., z_n$, to say that M is quasi-integral means that the z-coordinates of the points of M have a common denominator.

We can also reduce this to the notion of an integral point on a scheme. The following properties are equivalent:

(i) The set M is quasi-integral relative to A.

(ii) There is an A-scheme $\mathbf{X}$ of finite type such that,
(1) $X = \mathbf{X} \times_A K$,
(2) every point x of M extends to an A-valued point of $\mathbf{X}$.
(iii) There is an affine A-scheme $\mathbf{X}$ of finite type satisfying (1) and (2).

To show that (i) $\Rightarrow$ (iii), we choose an immersion, multiply the coordinates by a common denominator, and take the A-subalgebra generated by these.

To check that (ii) $\Rightarrow$ (i), we take $f \in \Lambda_X$; then there is $a \in A$, $a \neq 0$, such that af extends to $\mathbf{X}$; then the values of af at integral points are integers.

The implication (iii)$\Rightarrow$(ii) is trivial.

We now assume that X is an affine absolutely irreducible curve over K. We complete the normalisation of X and we then obtain the unique normal complete (projective) model $\bar{X}$ of X. We denote by $\bar{X}_\infty$ the points at infinity of $\bar{X}$; thus $\bar{X} - \bar{X}_\infty$ is the normalisation of X. We say that X *is exceptional* if $\bar{X}$ has genus 0 and $\bar{X}_\infty$ has at most two geometric (that is to say, after passing to an algebraic closure of K) points.

Siegel's Theorem. ([Si]). *If the affine curve X is not exceptional, every subset of $X(K)$ which is quasi-integral relative to A_S is finite.*

Here is the plan of the proof.
1). Review of the approximation of irrational numbers (essentially, the Thue-Siegel-Roth theorem)
2). Approximation of rational points on abelian varieties and on curves of genus ≥ 1.
3). Siegel's theorem for genus ≥ 1.
4). The case of genus 0 and at least 3 points at infinity.

7.2. Approximation of real numbers.

1). *For all real irrational numbers α, there is a sequence p_n/q_n of rational numbers with*

$$|\alpha - \frac{p_n}{q_n}| \leq \frac{1}{q_n^2}$$

Proof. Let Q be an integer > 1. We consider the numbers $q\alpha$, $1 \leq q \leq Q$. By the box principle, two of them, modulo 1, have difference $\leq 1/Q$. By subtracting, there is q, $1 \leq q \leq Q$, such that $||q\alpha|| \leq 1/Q$ (where $||\ ||$ is the distance to the nearest integer). In other words, there is $p \in \mathbf{Z}$ such that $|q\alpha - p| \leq 1/Q$. Therefore, $|\alpha - p/q| \leq \frac{1}{qQ} \leq \frac{1}{q^2}$.

As $\frac{1}{qQ} \to 0$ as $Q \to \infty$ we obtain then infinitely many such p_n/q_n.

One can improve this statement and replace $1/q_n^2$ by $1/\sqrt{5}q_n^2$, using continued fractions, and show that $\sqrt{5}$ is the best constant (see [HW, Theorems 193,194]).

We say that α is *approachable to within* $1/q^\delta$ if there is a sequence of rational numbers p_n/q_n, $q_n \geq 1$, with $|\alpha - p_n/q_n| \leq 1/q_n^\delta$.

2). *If $\delta > 2$, the set of real numbers α approachable to within $1/q^\delta$ has measure zero.*

Proof. We can limit ourselves to the segment [0,1]. For integers p and q with $0 \leq p \leq q$, $q > 0$, we denote by $I_{p,q}$ the interval

$$[\frac{p}{q} - \frac{1}{q^\delta}, \frac{p}{q} + \frac{1}{q^\delta}] \cap [0,1].$$

Let Q be an integer > 1 and let

$$I_Q = \bigcup_{q > Q, 0 \leq p \leq q} I_{p,q}.$$

Let $\alpha \in [0,1]$ and $\alpha \notin I_Q$, then α is not approachable to within $1/q^\delta$. But

$$\operatorname{meas}(I_Q) \leq O(\sum_{q \geq Q} \frac{q}{q^\delta}) = O(\sum_{q \geq Q} \frac{1}{q^{\delta-1}}).$$

As $\delta > 2$, the measure of I_Q tends to zero as Q tends to infinity; therefore the intersection of the I_Q, $Q > 1$, has measure zero.

The above result remains true with $1/q^\delta$, $\delta > 2$, replaced by $1/q\psi(q)$, where the series $\sum 1/\psi(q)$ converges.

3). **Roth's Theorem.** *If α is an irrational algebraic number, then for every $\delta > 2$, α is not approachable to within $1/q^\delta$.*

In other words, the set of pairs (p,q), $q \geq 1$, with $|\alpha - p/q| \leq 1/q^\delta$ is finite. But one does not know how to bound q as a function of α and δ. In contrast, the *number* of solutions can be effectively bounded (cf. [DR]).

Here is a more general statement for number fields. Let K be a number field. We choose a place v of K; we denote by $\alpha \mapsto |\alpha|_v$ the corresponding absolute value. There is a product formula on K compatible with the choice of v; let H be the corresponding height on K (cf.§§2.1,2.2).

Let $\alpha \in K_v$ be algebraic over K. Let $\delta > 2$. Then there is no sequence ω_n of distinct elements of K such that

$$|\alpha - \omega_n|_v \leq 1/H(\omega_n)^{\delta}.$$

For $K = \mathbf{Q}$ and v the archimedean absolute value, this is Roth's theorem. There is a more general statement using several places v and the product $\prod_v |\alpha_v - \omega_n|_v$ (cf. [L2],[L6],[Schm1],[Schm2],[Mi]). For the proof of Roth's theorem and its variants, we refer to Lang (*loc. cit.*).

Application to integral points.

Thue observed the following relation between diophantine equations and rational approximations. We take a sequence of solutions $(x, y) \in \mathbf{Z} \times \mathbf{Z}$, tending to infinity, of the equation $x^3 - 2y^3 = k$ ($k \in \mathbf{Z}$ fixed). Then $x/y \to \sqrt[3]{2}$. More precisely, $(x/y)^3 - 2 = k/y^3$, therefore $|x/y - \sqrt[3]{2}| \ll k/y^3$. Hence, an explicit bound on the approximations to $\sqrt[3]{2}$ provides an effective bound for the solutions of $x^3 - 2y^3 = k$. For example, Baker has shown that for all $p/q \in \mathbf{Q}$, $q > 0$, one has

$$|\sqrt[3]{2} - \frac{p}{q}| \geq \frac{10^{-6}}{q^{2.955}},$$

which implies that every solution $(x, y) \in \mathbf{Z} \times \mathbf{Z}$ of $x^3 - 2y^3 = k$ has

$$\sup(|x|, |y|) \leq (3.10^5.|k|)^{23}.$$

The exponent 2.955 of Baker has been improved to 2.43 by Chudnovsky [Cu]; so for the equation $x^3 - 2y^3 = k$, we obtain

$$\sup(|x|, |y|) \leq c.|k|^{1.76},$$

where c is effectively computable.

7.3. The approximation theorem on abelian varieties.

The argument above is insufficient for equations such as $y^2 - x^3 = k$. Siegel's method consists of refining the approximation theorem on abelian varieties.

Let V be an algebraic variety over K, immersed in projective space $\mathbf{P}_N$. Let K_v be a completion of K. We are interested in the approximation of algebraic points $\alpha \in V(K_v)$ by $\omega_n \in V(K)$, with $\omega_n \to \alpha$ in the v-adic topology.

To define a distance on the variety V at the point α, we choose a neighbourhood U of α with coordinates $z_1, ..., z_N$ taking the value 0 at α, and we define

$$d_v(\alpha, \beta) = \begin{cases} \sup_{1 \le i \le N} |z_i(\beta)|_v & \text{if } \beta \in U, \\ 1, & \text{if } \beta \notin U. \end{cases}$$

The logarithm of the distance is independent, up to a bounded function, of the coordinates chosen and the immersion of V in $\mathbf{P}_N$. (One could define in the same way, up to $O(1)$, the logarithm of the distance to a subvariety.)

Theorem. *If $\alpha \in V(K_v)$ is algebraic (i.e. $\alpha \in V(\bar{K})$) and if $\delta > 2$, there is no sequence of distinct elements $\omega_n \in V(K)$ such that $\omega_n \to \alpha$ and $d_v(\alpha, \omega_n) \le 1/H(\omega_n)^\delta$.*

This follows from Roth's theorem applied to the coordinates of ω_n.

When V is an abelian variety, this can be improved:

Theorem. *If V is an abelian variety, $\alpha \in V(K_v)$ is algebraic, and $\delta > 0$, then there is no sequence of distinct elements $\omega_n \in V(K)$ with $\omega_n \to \alpha$ and*

$$d_v(\alpha, \omega_n) \le \frac{1}{H(\omega_n)^\delta}.$$

Remarks. 1). This theorem is also true for any variety embeddable in an abelian variety, thus in particular for curves of genus ≥ 1 as they are embedded in their jacobians.
2). The theorem can be restated as:

$$\text{if } \omega_n \to \alpha, \text{ then } \log \frac{1}{d_v(\alpha, \omega_n)} = o(\log H(\omega_n)).$$

One can ask whether $o(\log H(\omega_n))$ may be replaced by a smaller estimate. This is indeed true. It has been proved by Masser, Wüstholz, Philippon, Waldschmidt ... that

$$\log \frac{1}{d_v(\alpha, \omega_n)} = O((\log \log H(\omega_n))^C),$$

for a suitable C (cf. [Ber]; it is not known whether C can be taken equal to 1). Moreover the constants involved are effective, provided a set of generators of the Mordell-Weil group of V over $K(\alpha)$ has been found.

Proof of the approximation theorem on abelian varieties.

We select a projective embedding of the abelian variety and we take the height H relative to this embedding.

Let $\delta > 0$ and let m be an integer ≥ 1 with $(m^2 - 1)\delta \geq 3$. Let (ω_n) be a sequence of distinct elements of $\Gamma = V(K)$ approaching α to within $1/H(\omega_n)^\delta$. From the weak Mordell-Weil theorem, $\Gamma/m\Gamma$ is finite. By taking a subsequence of (ω_n) we may assume that they are in the same class modulo $m\Gamma$: there are $\pi \in \Gamma$ and $\omega'_n \in \Gamma$ with

$$\omega_n = m\omega'_n + \pi.$$

As $V(K_v)$ is a compact group, again taking a subsequence we may assume that ω'_n converges to a point $\alpha' \in V(K_v)$. Then $\alpha = m\alpha' + \pi$, therefore α' is algebraic.

As the morphism $\omega \mapsto m\omega + \pi$ is étale, we have

$$d_v(\alpha', \omega'_n) \ll d_v(\alpha, \omega_n) \ll \frac{1}{H(\omega_n)^\delta}.$$

On the other hand, as $n \to \infty$,

$$\frac{\log H(\omega_n)}{\log H(\omega'_n)} \to m^2,$$

since $\log H$ is quadratic up to a bounded function (cf.§3.3); therefore for all sufficiently large n we have

$$H(\omega'_n)^{m^2-1} \leq H(\omega_n).$$

Then

$$d_v(\alpha', \omega'_n) \ll 1/H(\omega'_n)^{\delta(m^2-1)} \ll 1/H(\omega'_n)^3,$$

which contradicts the previous theorem since $3 > 2$.

Remark. 3). The approximation theorem on abelian varieties is ineffective because its proof uses Roth's theorem and the Mordell-Weil theorem, which are both ineffective at present. If the Mordell-Weil theorem could be made effective, one could dispense with Roth's theorem by using the abelian analogue of Baker's lower bounds for linear forms in logarithms, see Remark 2) above.

On the other hand, for the applications to the finiteness of integral points, work of Robinson-Roquette [Rob], and McIntyre [Mac] allows one to suppress the ineffectivity due to the Mordell-Weil theorem. They use a strong form of Roth's theorem, with several places, which allows the passage to a finite extension field (cf. §7.7). One can also obtain an algorithm giving the integral points on a curve, by assuming known a function $M(\alpha, \delta)$ ($\delta > 2$) such that all solutions $\omega \in V(K)$ of

$$d_v(\alpha, \omega) \leq 1/H(\omega)^\delta$$

satisfy $H(\omega) \leq M(\alpha, \delta)$. We sketch the idea.

Going back to the above proof, the problem consists now of finding π such that $\omega_n = m\omega'_n + \pi$. Consider the kernel V_m of multiplication by m, and write

$$H = \{z \in H^1(K, V_m) : \text{ the image of } z \text{ is zero in every } H^1(\text{local}, V)\}.$$

Then H is finite and effectively constructible. Let V_z be the image of z in $H^1(K, V)$; V_z is a principal homogeneous space, and it has points in all the completions, but one has no method for constructing rational points on V_z (this is the reason why the Mordell-Weil theorem is ineffective). But one can effectively construct a morphism $\phi : V_z \to V$ such that $\phi_z(x + a) = \phi_z(x) + ma$, $a \in V$ and $\phi_z(V_z(K)) = \{m\omega + \pi\}$ for a certain π. One can then dispense with π by choosing $\omega'_n \in V_z$ such that $\phi(\omega'_n) = \omega_n$. But V_z is not embedded in the same projective space as V. On V we have an ample line bundle L; the class of mL is invariant under the action of the Galois group. The obstruction to the existence of a sheaf defined over K on V_z corresponding to mL is given by an element of the Brauer group. But as there are points in all the completions, this element of the Brauer group is zero. To give an algorithm it is necessary to write the 2-cocycle corresponding to the element of the Brauer group as equal to a coboundary (all the cochains can be enumerated). Once one has the sheaf on V_z it provides a projective embedding and thus a height, and the proof can be completed as before.

[The fact that the Mordell-Weil theorem is not really necessary for the proof of Siegel (and Mahler) was already pointed out by Weil as early as 1935. See his Collected Papers, Vol. 1, p.99, where he writes:

"Siegel recourt, il est vrai, au théorème final de ma thèse, mais une analyse détaillée du mémoire montre que c'est inutile, et que le théorème des extensions non ramifiées suffit à la démonstration: de sorte que *les résultats de Siegel sont indépendants de la méthode de descente infinie.* Il en est de même, naturellement, des beaux résultats de Mahler, qui étendent notablement ceux de Siegel, au moins pour le genre 1."]

7.4. Application to curves of genus ≥ 1.

We apply the approximation theorem on an abelian variety to a curve embedded in its jacobian. Let X be a projective non-singular absolutely irreducible curve over K of genus ≥ 1. Let v be a place of K.

We consider a sequence of distinct points $P_n \in X(K)$; in particular their height tends to infinity. Let ϕ be a non-constant rational function on X defined over K. For sufficiently large n, P_n is not a pole of ϕ and $z_n = \phi(P_n)$ belongs to K.

Theorem. *We have*

$$\lim_{n\to\infty} \frac{\log |z_n|_v}{\log H(z_n)} = 0.$$

Proof. Assume this is false. By taking a subsequence and replacing ϕ by $1/\phi$ we may suppose that

$$\frac{\log |z_n|_v}{\log H(z_n)} \to \lambda, \text{ with } -\infty \leq \lambda < 0.$$

In particular, $z_n \to 0$ in K_v and by taking a subsequence we may assume that P_n converges to a zero P_0 of ϕ. As we are on a curve, P_0 is an algebraic point of X.

Between $H(P_n)$, the height corresponding to a morphism $X \to \mathbf{P}_N$, and $H(z_n)$, corresponding to a morphism $X \to \mathbf{P}_1$, we have an inequality (cf. §2.6)

$$H(z_n) \gg H(P_n)^{\text{const.}}.$$

On the other hand, if e is the multiplicity of P_0 as a zero of ϕ, we have $|z_n|_v \asymp d_v(P_n, P_0)^e$. Therefore, there is $c > 0$ such that

$$d_v(P_n, P_0) \leq 1/H(P_n)^c, \text{ for sufficiently large } n,$$

which contradicts the approximation theorem.

In the particular case $K = \mathbf{Q}$, $v =$ the infinite place, the preceding theorem can be equivalently stated:

$$\lim_{n\to\infty} \frac{\log |p_n|}{\log |q_n|} = 1,$$

where $z_n = p_n/q_n = \phi(P_n)$, $P_n \in X(\mathbf{Q})$. Therefore, the ratio of the number of digits in the numerators and denominators of rational points tends to 1, which is more precise than the finiteness of the number of integral points.

7.5. Proof of Siegel's theorem.

Let X be a curve of genus ≥ 1 defined over a number field K, ϕ a non-constant rational function on X defined over K, P_n a sequence of distinct points of $X(K)$, v a place of K. We have proved that (§7.4)

$$\frac{\log |z_n|_v}{\log H(z_n)} \to 0, \tag{1}$$

where $z_n = \phi(P_n)$.

We shall deduce the following result: if S is a finite set of places of K containing the archimedian places S_∞, and if A_S is the ring of S-integers of K, then there are only finitely many n such that $\phi(P_n) = z_n \in A_S$.

If not, by taking a subsequence, we have $z_n \in A_S$ for all n, that is to say, $|z_n|_w \leq 1$ for $w \notin S$ and for all n. But the height $H(z)$ has been defined as

$$H(z) = \prod_{v \in M_K} \sup(1, |z|_v),$$

where the $|.|$ are normalised to satisfy the product formula (§2.1). Thus for $z \in A_S$

$$H(z) = \prod_{v \in S} \sup(1, |z|_v),$$

and

$$\log H(z) = \sum_{v \in S} \sup(0, \log |z|_v).$$

By applying this to z_n we get

$$1 = \sum_{v \in S} \sup(0, \frac{\log |z_n|_v}{\log H(z_n)}) \leq \sum_{v \in S} \frac{|\log |z_n|_v|}{\log H(z_n)},$$

which contradicts formula (1) as $n \to \infty$.

Corollary. *If X is an affine curve defined over K of genus ≥ 1, every set of K-rational points of X which is quasi-integral relative to A_S is finite.*

It remains to treat the case of curves of genus 0 with m points at infinity, $m \geq 3$. By making an extension of scalars, we can assume that these points are rational. We shall reduce to the case of 3 points at infinity.

If $\phi : X \to Y$ is a morphism of affine varieties, over K, and if $\Sigma \subset X(K)$ is quasi-integral relative to A_S, then $\phi(\Sigma)$ is also quasi-integral. In particular, if we remove one point Q from X, we have a morphism $X - \{Q\} \to X$. If $Q_1, ..., Q_m$ are the m points at infinity of $\bar{X}$, by considering the morphism $\bar{X} - \{Q_1, ..., Q_m\} \to \bar{X} - \{Q_1, Q_2, Q_3\}$ we are reduced to the case of 3 points at infinity.

When $m = 3$ we can assume that $X = \mathbf{P}_1 - \{0, 1, \infty\}$. The affine ring of $\mathbf{P}_1 - \{\infty\}$ is $K[T]$, and the affine ring of X is $K[T, 1/T, 1/(T-1)] = K[T, 1/T(T-1)]$. If we write $t = T$, $x = 1/T$, $y = 1 - T$, $z = 1/(1-T)$ we obtain another presentation of the affine ring of X, with generators t, x, y, z and relations

$$tx = 1, yz = 1, y + t = 1. \tag{2}$$

If Y is an affine variety, a morphism $Y \to X$ amounts to a morphism of the ring $K[T, 1/T, 1/(T-1)]$ of X to the ring Λ_Y of Y, that is to say elements t, x, y, z of Λ_Y satisfying (2); the relations (2) mean that y and t are units of sum 1. (The units of a number field of sum 1 have been studied since Nagell under the name of exceptional units.)

Let Σ be a quasi-integral subset of $X(K)$. On Σ the functions $t, 1/t, 1/(t-1)$ take their values in $D^{-1}A_S$, for some $D \in K$, $D \neq 0$. We identify the points of Σ with their corresponding values of t.

We first show that $t = ut_0$, where t_0 belongs to a finite set, and $u \in A_S^*$. This amounts to proving that the A_S-ideals generated by the values of t are finite in number. But

$$t = D^{-1}\alpha, \alpha \in A_S, \text{ and } \frac{1}{t} = D^{-1}\beta, \beta \in A_S,$$

therefore

$$\alpha\beta = D^2 \text{ and } (\alpha) \text{ divides } (D^2).$$

In the same way $y = 1 - t = vy_0$, where y_0 belongs to a finite set and v is a unit of A_S. Therefore $t_0u + y_0v = 1$, and clearing the denominators we reduce to a finite number of equations

$$Au + Bv = C, \tag{3}$$

where A,B,C are given non-zero integers in A_S and the unknowns u,v are units of A_S.

Siegel's theorem in this particular case states that there are only a finite number of $u, v \in A_S^*$ satisfying such an equation.

[In particular the equation $u + v = 1$ defining exceptional units is of this type.]

There are several ways to complete the proof. Here is Siegel's method which consists of reducing to a curve of higher genus.

We suppose that there are infinitely many solutions (u_n, v_n) of equation (3). We know that the group Γ of S-units is finitely generated. Let m be an integer ≥ 3; then Γ/Γ^m is finite. By taking a subsequence we can assume that

$$u_n = U_n^m u_0, v_n = V_n^m v_0, (n \geq 1),$$

where $u_0, v_0 \in \Gamma$ are fixed elements (independent of n). Putting $\alpha = Au_0$, $\beta = Bv_0$, $\gamma = C$, we obtain infinitely many quasi-integral points (U_n, V_n) on the curve

$$\alpha X^m + \beta Y^m = \gamma.$$

As $m \geq 3$ the genus of this curve is ≥ 1, so we have a contradiction, which finishes the proof.

(For the equation $\alpha X^m + \beta Y^m = \gamma$, one can use directly the result of Thue.)

Remarks.

1). *On Siegel's proof.*

In 1929, Siegel had only a weaker form of the approximation theorem of Roth (1955): in place of the exponent $2 + \epsilon$, Siegel had the exponent $2\sqrt{n}$, where n is the degree of $\mathbf{Q}(\alpha)$ over $\mathbf{Q}$. But in the proof, the field $\mathbf{Q}(\alpha)$ may have to be extended (one writes $\alpha = m\alpha' + \pi$, where α' is algebraic over $\mathbf{Q}(\alpha)$). In the case of genus 1, enough is gained from the height to compensate this degree increase. This is not true for genus ≥ 2 and Siegel then had to use simultaneous approximations. But all this part of Siegel's proof can now be elided, thanks to Roth's theorem.

2). *S-integral points.*

Siegel proved his theorem for the usual integral points ($S = S_\infty$). In 1933, Mahler extended this result to S-integral points, but only in the case of genus 1 (because of the above difficulties) and over the rationals $\mathbf{Q}$. He used his p-adic analogue of Siegel's approximation theorem.

In 1955, Roth proved his theorem with the exponent $2+\epsilon$ for an archimedian valuation. In 1958, Ridout treated the case of an arbitrary valuation. Some generalisations were given by Lang [L7] and LeVêque [Le] around 1960. Lang showed that Roth's theorem could be extended, under reasonable hypotheses, to any field equipped with a product formula, in particular to function fields in characteristic 0.

(If ϕ is a formal power series in one variable T over $\mathbf{C}$,. there is a sequence f_n/g_n of rational functions, where f_n, g_n are polynomials of degree $\leq d_n$, such that $\phi - f_n/g_n$ begins with terms in T^{2d_n}. But if ϕ is algebraic over $\mathbf{C}(T)$, and if κ is a real number > 2, the analogue of Roth's theorem is that there is no sequence with $\phi - f_n/g_n = O(T^{\kappa d_n})$. Such results had already been obtained by Fenna [Fe] and Uchiyama [U] between 1956 and 1960.)

Siegel's theorem is then valid for all rings (of characteristic 0) of finite type over $\mathbf{Z}$.

Roth's theorem for function fields does not extend directly to characteristic p (consider $\phi = T + T^p + T^{p^2} \ldots$ which is a root of the Artin-Schreier polynomial $X^p - X + T = 0$). It would be interesting to know the analogue in characteristic p of Roth's theorem and of Siegel's theorem. The only counterexamples known arise from the analogue of Mordell's conjecture over functions fields and are obtained by composing a constant map with the Frobenius endomorphism (cf. §5.7). Further, it would be interesting to combine Siegel's arguments with those of Mumford saying that the rational points are widely spaced (§5.7).

7.6. Application to $P(f(n))$.

For an integer $n \neq 0, \pm 1$, we denote the greatest prime factor of n by $P(n)$.

Theorem. (Siegel). *Let $f \in \mathbf{Z}[X]$ be a non-constant polynomial having at least two distinct zeros over $\mathbf{C}$. Then $P(f(n)) \to \infty$ with n.*

For example, for every sufficiently large integer n, $n^3 + 2$ has a prime factor $\geq 10^{10}$.

First proof.

If not, there is a sequence (n_i) of integers tending to infinity, and an integer M such that $|f(n_i)| > 1$ for all i, and all the prime factors of $f(n_i)$ are bounded by M. If S is the set of prime numbers $\leq M$, the $1/f(n_i)$ are S-integers. But $\mathbf{Q}[T, 1/f(T)]$ is the affine algebra of the variety obtained by removing the zeros of f from the affine line; this variety is an affine curve of genus 0, where the point at infinity and at least two other points (the zeros of f) have been removed. From Siegel's theorem, the set of S-integral points is finite.

Second proof.

As Siegel did not have the finiteness theorem for arbitrary S, he proceeded differently. If the $f(n_i)$ involve only the prime numbers $p_1, ..., p_m$, we can write

$$f(n_i) = \pm p_1^{h_1} ... p_m^{h_m}.$$

Let l be an integer $> \deg f$. We reduce the h_j modulo l and obtain:

$$f(n_i) = \pm p_1^{\alpha_1} ... p_m^{\alpha_m} y_i^l, \text{ with } 0 \leq \alpha_i < l,$$

for some $y_i \in \mathbf{Z}$. But each of the curves

$$f(X) = \pm p_1^{\alpha_1} ... p_m^{\alpha_m} Y^l,$$

is irreducible, and of genus ≥ 1 (because of the choice of l) and has only finitely many integral points.

Both proofs are ineffective. It is now known by effective methods that

$$P(f(n)) \gg_f \log \log n.$$

(see [BM, Chapter 3]).

7.7. Effectivity.

The above exposition is based on a version of Roth's theorem for a single absolute value. Here is a statement for a finite number of places.

Let S be a finite set of places, and let ω_n be a sequence of elements of K, with $H(\omega_n) \to \infty$. For each $v \in S$ we suppose given $\alpha_v \in K_v$, α_v algebraic over K. If δ is a real number > 2, then one cannot have

$$\prod_{v \in S} |\omega_n - \alpha_v|_v \leq \frac{1}{H(\omega_n)^\delta}, \text{ for all } n.$$

(see [L2]; some previous results are due to Mahler, Parry).

Using this, the approximation theorem on abelian varieties (§7.3) can be proved in the form: if one takes $\alpha_v \in V(K_v)$, α_v algebraic, for each

$v \in S$, and if $\delta > 0$, then there is no sequence ω_n of distinct elements of $V(K)$ with

$$\prod_{v \in S} d(\omega_n, \alpha_v)_v \leq \frac{1}{H(\omega_n)^\delta}.$$

This is true by Roth's theorem if $\delta > 2$ (by projecting onto a line). On the other hand, to prove this formula K can be replaced by a finite extension L, under the condition that S be replaced by the set of places of L above those of S: indeed, for $\lambda \in K_v$, we have

$$|\lambda|_v = \prod_{w|v} |\lambda|_w.$$

Let M be an integer such that $(m^2 - 1)\delta \geq 3$. We replace the ω_n by their mth roots: $m\omega'_n = \omega_n$. From the proof of the Mordell-Weil theorem, there is an extension L of K in which these ω_n are mth roots (L is a finite extension unramified outside m and the places of bad reduction). Using this version of Roth's theorem, the ineffectivity due to the Mordell-Weil theorem is removed.

8. BAKER'S METHOD

Baker's method does not actually prove Siegel's theorem in full generality, but in every case where it applies, it is effective.

8.1. Reduction theorems.

We first study the relations between quasi-integral points under morphisms of affine varieties.

We keep the notation of the previous chapter: K is a number field, $A = A_S$ the ring of S-integers of K (most of the following extends to any integral domain A).

1). *If $\phi : X \to Y$ is a morphism of affine varieties, and if $\Sigma \subset X(K)$ is a set of rational points, quasi-integral relative to A, then $\phi(\Sigma)$ in $Y(K)$ is quasi-integral relative to A.*

This is clear.

2). A morphism $\phi : X \to Y$ is *finite* if and only if it is proper and has finite fibres. If X and Y are affine with affine rings Λ_X and Λ_Y, this means that Λ_X is a finite Λ_Y-module, or that Λ_X is integral over Λ_Y.

If $\phi : X \to Y$ is a finite morphism of affine varieties, and if Σ is a subset of $X(K)$, then Σ is quasi-integral if and only if $\phi(\Sigma)$ is quasi-integral.

Proof.

Suppose that $\phi(\Sigma)$ is quasi-integral. Let $f \in \Lambda_X$. As ϕ is finite, f is integral over Λ_Y: there is a relation

$$f^m + g_1 f^{m-1} + ... + g_m = 0,$$

where the g_i belong to the image of Λ_Y in Λ_X. There is $D \in A$, $D \neq 0$, such that

$$g_i(\Sigma) \in \frac{1}{D}A, \text{ for } i = 1, ..., m.$$

Then

$$(Df)^m + Dg_1(Df)^{m-1} + ... + D^m g_m = 0.$$

Thus for $\sigma \in \Sigma$, $Df(\sigma)$ is an element of K integral over A. As A is integrally closed, we have $Df(\sigma) \in A$.

Consequences. a). To study quasi-integral sets, we may assume that the curve is non-singular (use the normalisation map which is finite).

b). Let X be a non-singular affine curve, $\bar{X}$ its projective completion, and $X_\infty = \bar{X} - X$. From the Riemann-Roch theorem, there is a function $\phi : \bar{X} \to \mathbf{P}_1$ whose set of poles is X_∞; then $\phi|_X$ gives a finite morphism $X \to \mathbf{P}_1 - \{\infty\}$, and a subset Σ of $X(K)$ is quasi-integral if and only if its image is quasi-integral. Thus finiteness theorems can be reformulated in terms of functions on the curve.

Assertion 2) only gives us information on the points of Y which are liftable to X over K. We need more.

3). *Let $\phi : X \to Y$ be a morphism of affine varieties; assume that ϕ is finite, étale, and surjective. If Σ_Y is a quasi-integral set of points of Y, there is a finite extension K' of K such that $\phi^{-1}(\Sigma_Y)$ is rational over K'.*

We first show that there is $\Delta \in A$, $\Delta \neq 0$ such that if $\sigma \in \Sigma_Y$, then $K(\phi^{-1}(\sigma))$ is an extension of K unramified outside Δ.

(If for example we take X and Y to be the affine line, and the morphism ϕ to be $z \mapsto z^2$, then ϕ is finite and étale outside 0, but is not étale at 0, and the field $K(\sqrt{\alpha})$, $\alpha \in A$ can be ramified anywhere. But if we take X and Y to be the affine line with 0 removed, and $\phi : z \mapsto z^2$, then ϕ is finite and étale, and an integral point is a unit α of K and then $K(\sqrt{\alpha})$ is unramified outside the divisors of 2.)

The proof of the existence of Δ is the same as that of the Chevalley-Weil theorem (which assumes that the varieties are projective and the morphisms are everywhere étale; here the morphisms can be ramified at infinity cf. §4.2).

For $\Delta \in A$, $\Delta \neq 0$, we consider the ring $A_\Delta = A[\frac{1}{\Delta}]$. By elementary arguments of algebraic geometry (cf. [G, IV,§8]) one can find affine schemes $\mathbf{X}$, $\mathbf{Y}$ of finite type over a suitable A_Δ such that

$$\mathbf{X} \times_A K = X, \mathbf{Y} \times_A K = Y,$$

and a finite étale surjective morphism $\Phi : \mathbf{X} \to \mathbf{Y}$ corresponding to ϕ. The elements of Σ_Y are points of $X(K)$ having a common denominator, and we can assume that $\Sigma_Y \subset \mathbf{Y}(A_\Delta)$. A point σ of Σ_Y can be viewed as a morphism Spec $A_\Delta \to \mathbf{Y}$. Then σ lifts over an extension of K unramified outside the divisors of Δ.

The proof of 3) is now completed using Hermite's theorem (cf.§4.1): there are only finitely many extensions of bounded degree and unramified outside a given finite set of places. Here the degree is bounded by the number of points in the fibres of ϕ.

Exercise. Let $\phi : X \to Y$ be a finite étale surjective morphism of non-singular affine curves over a field of characteristic 0. Then X is exceptional (that is, of genus 0 with at most two points at infinity) if and only if Y is exceptional.

[Note that the finite unramified coverings of $\mathbf{P}_1 - \{0, \infty\}$ are given by $z \mapsto z^n$, $n \in \mathbf{Z}$.]

4). If X is a variety over K, let $(F_{X,K})$ denote the property:
$(F_{X,K})$ Every set of quasi-integral points of $X(K)$ is finite.

Let (F_X) denote the property:
(F_X) We have $(F_{X,K'})$ for every finite extension K' of K.

Proposition.

a). *Let $\phi : X \to Y$ be a morphism with finite fibres. Then $(F_{Y,K}) \Rightarrow (F_{X,K})$.*

b). *Let $\phi : X \to Y$ be a finite étale surjective morphism. Then (F_X) is equivalent to (F_Y).*

Proof.

a). If $(F_{Y,K})$ is true, and if Σ_X is a quasi-integral subset of $X(K)$, then $\phi(\Sigma_X)$ is a quasi-integral subset of $Y(K)$, therefore finite, and as the fibres are finite Σ_X is finite.
b). It is clear that $(F_Y) \Rightarrow (F_X)$, from a). We show that $(F_X) \Rightarrow (F_Y)$. Let Σ_Y be a quasi-integral subset of $Y(K)$. From reduction 3), we can assume (after enlarging K) that the points of Σ_Y lift: there is $\Sigma_X \subset X(K)$ such that $\phi(\Sigma_X) = \Sigma_Y$. From the reduction 2), Σ_X is a quasi-integral subset. Then (F_X) implies that Σ_X is finite, thus its image by ϕ is finite.

Remarks. 1). All the reduction theorems above are effective.
2). Siegel's trick of reducing the non-exceptional case of genus 0 to the case of genus ≥ 1 can now be explained by considering the Fermat curve $X^l + Y^l = Z^l$ (which has genus ≥ 1 for $l \geq 3$) as an étale covering of $\mathbf{P}_1 - \{0, 1, \infty\}$, via the map $(X, Y, Z) \mapsto (X/Z)^l$. Hence to prove the finiteness theorem for $\mathbf{P}_1 - \{0, 1, \infty\}$ amounts to proving it for the Fermat curve.

8.2. Lower bounds for $\sum \beta_i \log \alpha_i$.

Baker's method provides lower bounds for non-zero linear combinations $\sum \beta_i \log \alpha_i$, where α_i and β_i are algebraic numbers. When the numbers β_i are ordinary integers (we denote these by b_i) this is equivalent to bounding $(\prod \alpha_i^{b_i}) - 1$ from below.

Let K be a number field of degree $\leq D$ over $\mathbf{Q}$, $\alpha_1, ..., \alpha_n$ some elements of K, $b_1, ..., b_n$ rational integers, and v an absolute value of K. The bounds can be written in terms of

$$B = \sup(4, |b_1|, ..., |b_n|),$$

and the heights of the α_i. Baker uses the following notion of height:

Let α be an algebraic number of degree $d(\alpha) = d$; we can write a minimal equation of α in the form

$$a_0\alpha^d + a_1\alpha^{d-1} + ... + a_d = 0,$$

where $a_0, ..., a_d$ are rational integers of greatest common divisor 1. We define

$$A(\alpha) = \sup |a_i|.$$

For example, if $\alpha = p/q$ is a rational number with $(p,q) = 1$, then $A(\alpha)$ is the usual height $\sup(|p|,|q|) = H(p/q)$. In the general case, it is easily proved that

$$A(\alpha) \asymp_{d(\alpha)} H(\alpha)^{d(\alpha)},$$

where $H(\alpha)$ is the normalised height of α. Geometrically, one sees this by considering $(a_0, ..., a_d)$ as a point of projective space $\mathbf{P}_d$; on the other hand, α and its conjugates give a point in the d-fold symmetric product of the projective line. But $\mathbf{P}_1 \times ... \times \mathbf{P}_1/S_d$ is naturally isomorphic to $\mathbf{P}_d$, and the above equivalence follows from this.

Explicitly, we have

$$A(\alpha) \leq 2^{d(\alpha)} H(\alpha)^{d(\alpha)}, \text{ and } H(\alpha)^{d(\alpha)} \leq (1 + d(\alpha))A(\alpha).$$

It is possible to replace $2^{d(\alpha)}$ by $\binom{d(\alpha)}{[d(\alpha)/2]}$ and $1 + d(\alpha)$ by $1 + d(\alpha)^{1/2}$.

Put $A_i = \sup(4, A(\alpha_i))$, $1 \leq i \leq n$.

Baker's Theorem. (cf. [Ba], [BM], [L5]). *If* $\prod_{i=1}^n \alpha_i^{b_i} \neq 1$, *then*

$$|(\prod_{i=1}^n \alpha_i^{b_i}) - 1|_v > \exp\{-c\Omega(\log\Omega)(\log B)\},$$

where $\Omega = \prod_{i=1}^n \log A_i$, *and* c *is an effectively computable constant which depends only on* K_v,D *and* n.

For example, if $K_v = \mathbf{C}$, we can take c to be the number $(16nD)^{200n}$; this can be refined to $2^{12n+53}n^{2n}D^{n+2}$ (cf. [Wa2]).

We shall use this theorem in the following form.

Corollary. *If* $\prod_{i=1}^n \alpha_i^{b_i} \neq 1$, *then*

$$|(\prod_{i=1}^{n} \alpha_i^{b_i}) - 1|_v > 1/B^{c(\alpha)},$$

where $c(\alpha)$ *is a constant depending on* $K_v, \alpha_1, ..., \alpha_n$.

For every algebraic number $x \neq 0$, the product formula implies the inequality $|x|_v \geq 1/H(x)$. Applied to $x = (\prod_{i=1}^n \alpha_i^{b_i}) - 1$ this gives

$$|(\prod_{i=1}^{n} \alpha_i^{b_i}) - 1|_v \geq 1/c^B,$$

where c depends only on $\alpha_1, ..., \alpha_n$ and K_v. In place of c^B, it is essential for the applications to have a function which increases less quickly than an exponential of B. Baker's first results, with a lower bound of the form $\exp\{-c(\log B)^\kappa\}$, amply sufficed. Nevertheless, the box principle shows that the correct order of magnitude is indeed $1/B^{c(\alpha)}$.

For the most general linear combination $\sum \beta_i \log \alpha_i$ with the β_i algebraic, one has an analogous assertion with $\Omega(\log \Omega)(\log B\Omega)$, where B is an upper bound for the heights of the β_i (see [BM, Chapters 1 and 2]).

8.3. Application to $\mathbf{P}_1 - \{0, 1, \infty\}$.

As usual, we denote by A_S the ring of S-integers of K. To show the finiteness of quasi-integral sets of $\mathbf{P}_1 - \{0, 1, \infty\}$ amounts to showing that if A, B, C are non-zero elements of A_S, then the equation

$$Au + Bv = C,$$

has only finitely many solutions u, v in A_S^* (cf. §7.5, Equation (3)).

As the group A_S^* of S-units is finitely generated, it can be written $\mu \times L$ where μ is a finite group (group of roots of unity in K) and L is a free abelian group. We select a basis $\epsilon_1, ..., \epsilon_r$ of L, so that an S-unit $u \in A_S^*$ can be written uniquely

$$u = \zeta \epsilon_1^{b_1} ... \epsilon_r^{b_r},$$

with ζ a root of unity and $b_i \in \mathbf{Z}$. We then put $b(u) = \sup_{1 \leq i \leq r} |b_i|$.

That u is an S-unit means that $|u|_w = 1$ if $w \notin S$. If we associate to $u \in A_S^*$ the family $(\log |u|_w)_{w \in S}$ of logarithms of its absolute values at the places of S, we obtain an embedding

$$l : A_S^*/\mu = L \hookrightarrow \mathbf{R}^{|S|}.$$

The product formula implies

$$\sum_{w \in S} \log |u|_w = 0,$$

so the image of L is a discrete subgroup contained in a hyperplane of $\mathbf{R}^{|S|}$. On this discrete subgroup there are two equivalent norms one given by $\sup |b_i|$, the other by the coordinates of $\mathbf{R}^{|S|}$.

Hence there is a constant $c > 0$ such that for all $u \in A_S^*$ we have

$$c^{-1} \sup_{w \in S} |\log |u|_w| \leq b(u) \leq c \sup_{w \in S} |\log |u|_w|.$$

It is convenient to remove the absolute values of the logarithms, which can be done by means of the following remark: if x_w, $(w \in S)$, are real numbers with $\sum_{w \in S} x_w = 0$, then $\sup_{w \in S} |x_w| \leq (|S| - 1) \sup x_w$. We thus obtain:

Lemma. *There is a constant $c > 0$ such that for all $u \in A_S^*$ we have*

$$c^{-1} \sup_{w \in S} \log |u|_w \leq b(u) \leq c \sup_{w \in S} \log |u|_w.$$

Suppose now there is an infinite sequence (u_n, v_n) of solutions of the equation $Au + Bv = C$. By taking a subsequence, we may assume (by symmetry) that $b(u_n) \geq b(v_n)$, and that there is $w \in S$ with $\log |u_n|_w \geq b(u_n)/c$. Writing $u = u_n$, $v = v_n$, we then have

$$|-\frac{Bv}{Au} - 1|_w < c_1 c_2^{-b(u)},$$

where $c_1 > 0$ and $c_2 > 1$ are independent of u. We write

$$u = \zeta \epsilon_1^{b_1} \ldots \epsilon_r^{b_r}, v = \zeta' \epsilon_1^{b'_1} \ldots \epsilon_r^{b'_r}, \alpha = -\frac{B}{A}\frac{\zeta'}{\zeta},$$

and we have

$$|\epsilon_1^{b'_1 - b_1} \ldots \epsilon_r^{b'_r - b_r} \alpha - 1|_w < c_1 c_2^{-b(u)}.$$

As $Au + Bv = C \neq 0$, the left hand side is non-zero, and Baker's theorem gives

$$|\epsilon_1^{b'_1 - b_1} \ldots \epsilon_r^{b'_r - b_r} \alpha - 1|_w > 1/b(u)^{c_3}.$$

As c_1,c_2,c_3 are independent of n, we obtain the finiteness of the number of solutions.

This method gives an effective bound for the solutions of $Au+Bv=C$, knowing $\epsilon_1, ..., \epsilon_r$. On the other hand, one can also find effectively a system of generators $\epsilon_1, ..., \epsilon_r$ of L, with an explicit bound for the absolute values of the ϵ_i ([Gy], [Si2]).

8.4. Applications to other curves.

We have dealt with the crucial case of the projective line minus 3 points. To go further, we use the following two reduction principles (see §8.1(4)): Let X and Y be two affine curves, $f: X \to Y$ a morphism with finite fibres. Then:

Reduction 1. We have $(F_Y) \Rightarrow (F_X)$.

Reduction 2. If f is finite, étale and surjective, then $(F_X) \Leftrightarrow (F_Y)$.

If $\bar{X}$ and $\bar{Y}$ are the corresponding projective curves, we can write $X = \bar{X} - X_\infty$, $Y = \bar{Y} - Y_\infty$, and f extends to a morphism $\bar{f}: \bar{X} \to \bar{Y}$ such that $\bar{f}^{-1}(Y_\infty) \subset X_\infty$. For f to be proper (equivalently, finite), it is necessary and sufficient that $\bar{f}^{-1}(Y_\infty) = X_\infty$, and then $Y_\infty = \bar{f}(X_\infty)$. For the second reduction one further needs that the covering $\bar{X} \to \bar{Y}$ be unramified outside Y_∞.

Case 1. *Thue's equation* $F(x,y)=m$.

Let F be a homogeneous polynomial in X and Y with coefficients in the number field K, having at least 3 distinct linear factors over the algebraic closure. Let $m \in K$, $m \neq 0$. Then the equation $F(x,y) = m$ has only a finite number of solutions x, y in A_S.

To see this, we apply the first reduction principle to the morphism which sends the point (x,y) of the affine curve X (with equation $F(x,y) = m$) to the line which joins this point to the origin:

$$X \to \mathbf{P}_1 - \{\text{zeros of the polynomial } F(1,t)\}$$

$$(x,y) \mapsto t = y/x.$$

As at least 3 points have been removed from $\mathbf{P}_1$, we obtain the result we want.

[When Baker applies his theorem to the effective approximation of an algebraic number α by rational numbers p/q, he selects two conjugates α' and α'', constructs a morphism to $\mathbf{P}_1 - \{\alpha, \alpha', \alpha''\}$, and changes the variables to send the 3 points to $0, 1, \infty$. He thus reduces the problem to an equation $Au + Bv = C$.

This method gives a bound for the absolute values of x, y. For example, over $\mathbf{Z}$, one has a bound

$$\max(|x|, |y|) \leq (m+1)^{C(F)},$$

for the solutions of $F(x,y) = m$.]

Case 2. *Elliptic curves.*

Let X be an affine curve whose compactification $\bar{X}$ is an elliptic curve. As X is affine, $X_\infty = \bar{X} - X$ is non-empty. By extending the ground field it suffices to treat the case where X_∞ has exactly one point, rational over K. We take this point as origin and put on $\bar{X}$ the corresponding group law. In classical language, this amounts to saying that we study the Weierstrass equation $y^2 = f(x)$ where f is a polynomial of degree 3.

Multiplication by 2 is étale and the inverse image of 0 consists of 4 points $0, E_1, E_2, E_3$ (the subgroup of 2-division points). By the second reduction, proving the theorem for $\bar{X} - \{0\}$ reduces to proving it for $\bar{X} - \{0, E_1, E_2, E_3\}$. We then apply the first reduction to the morphism given by the coordinate x:

$$x : \bar{X} - \{0, E_1, E_2, E_3\} \to \mathbf{P}_1 - \{\infty, e_1, e_2, e_3\},$$

where $f(x) = (x - e_1)(x - e_2)(x - e_3)$. As the finiteness result is true for $\mathbf{P}_1 - \{\infty, e_1, e_2, e_3\}$, it is then true for X.

Reductions of this type are quite old. They have been used by Siegel, Chabauty, Kubert, Lang and others.

Case 3. *Hyperelliptic case.*

Assume that $\bar{X}$ is hyperelliptic of genus $g \geq 2$. Let σ be the hyperelliptic involution of $\bar{X}$, so that $\bar{X}/\{1, \sigma\}$ is isomorphic to $\mathbf{P}_1$ over some extension of K.

Assume further that $X_\infty \cap \sigma X_\infty \neq \emptyset$. By the first reduction we may assume (after some field extension) that X_∞ is:

either a rational point P, fixed under σ,

or a pair $(P, \sigma P)$, where P is rational and not fixed by σ.

This means that X is the affine curve defined by the equation

$$y^2 = f(x),$$

where f is a monic polynomial of degree $2g+1$ or $2g+2$ without multiple roots.

Let $\alpha_1, \alpha_2, \alpha_3$ be three distinct roots of f. We introduce the elliptic curve E defined by the equation

$$t^2 = (x - \alpha_1)(x - \alpha_2)(x - \alpha_3).$$

Then via the map x, $\bar{E}$ is a covering of the projective line, ramified at $\alpha_1, \alpha_2, \alpha_3$ and ∞. On the other hand, the map $\bar{X} \to \mathbf{P}_1$, defined by $(x, y) \mapsto x$, is ramified at $\alpha_1, \alpha_2, \alpha_3$ and some other points:

$$\begin{array}{ccc} \bar{E} & & \bar{X} \\ & \searrow \quad \swarrow & \\ & \mathbf{P}_1 & \end{array}$$

Consider the normal projective curve $\bar{Z}$, whose function field is the composite of the function fields of $\bar{E}$ and $\bar{X}$; if x, y are variables for $\bar{X}$, and t, x variables for $\bar{E}$, then x, y, t are variables for $\bar{Z}$. We denote by Z_∞ the points of $\bar{Z}$ above $\infty \in \mathbf{P}_1$, and by X_∞, E_∞ those of $\bar{X}$, $\bar{E}$:

$$\begin{array}{ccc} & \bar{Z} & \\ & \swarrow \quad \searrow & \\ \bar{E} & & \bar{X} \\ & \searrow \quad \swarrow & \\ & \mathbf{P}_1 & \end{array}$$

The finiteness theorem is true for $\bar{E} - E_\infty$, therefore by the first reduction it is true for $\bar{Z} - Z_\infty$. To apply the second reduction, it suffices to check that the arrow $\bar{Z} \to \bar{X}$ is proper and étale. By base change it is clearly proper. To see that it is étale, we use the fact that $\bar{E} \to \mathbf{P}_1$ is ramified only at $\alpha_1, \alpha_2, \alpha_3$, and ∞, and $\bar{X} \to \mathbf{P}_1$ is ramified at $\alpha_1, \alpha_2, \alpha_3$ and these ramifications cancel each other (Abhyankar's lemma). Equivalently, to pass to $\bar{E}$ we extract the square root of $(x-\alpha_1)(x-\alpha_2)(x-\alpha_3)$; to pass to $\bar{X}$ we extract the square root of $(x-\alpha_1)...(x-\alpha_m)$; thus for $\bar{Z} \to \bar{X}$ it suffices to extract the square root of $(x - \alpha_4)...(x - \alpha_m)$, which gives no ramification at $\alpha_1, \alpha_2, \alpha_3$. Thus the finiteness theorem is true for X.

The first case not treated by Baker's method is an hyperelliptic curve of genus ≥ 2 with $X_\infty = \{P\}$ and $P \neq \sigma P$. For example, we take $y^2 = f(x)$ with f of degree 6 and add one of the two points at infinity. Siegel's theorem gives the finiteness of integral points, but whether it can be made effective is unclear.

Case 4. *Superelliptic equations.*

We consider the equation $y^m = f(x)$, where m is an integer ≥ 3, and f has at least 2 simple roots:

$$f(x) = (x-\alpha)(x-\beta)g(x),\ \alpha \neq \beta, g(\alpha) \neq 0, g(\beta) \neq 0.$$

We introduce the curve E defined by $t^m = (x-\alpha)(x-\beta)$; E is hyperelliptic (or elliptic if $m = 3$ or 4):

$$(x - \frac{\alpha+\beta}{2})^2 = t^m + \gamma,\ \gamma = \frac{(\alpha+\beta)^2}{4} - \alpha\beta.$$

We use the same principle:

$$\begin{array}{ccccc} & & \bar{Z}\ (t,x,y) & & \\ & \swarrow & & \searrow & \\ (t,x)\ \bar{E} & & & & \bar{X}\ (x,y) \\ & \searrow & & \swarrow & \\ & & \mathbf{P}_1 & & \end{array}$$

The finiteness theorem is true for E, thus for Z, and $\bar{Z} \to \bar{X}$ is étale outside ∞ by the same arguments as before.

Remark. A further example of an effective (and completely explicit) result obtained by Baker's method is the following [L5, Appendix to Chapter 6] :

Theorem. *Let $f \in \mathbf{Z}[x,y]$ be an absolutely irreducible polynomial for which the curve $f(x,y) =$ has genus 1, and let H be the maximum of the (usual) absolute values of the coefficients of f. Then for every point $(x,y) \in \mathbf{Z} \times \mathbf{Z}$ on the curve, we have*

$$\sup(|x|,|y|) \leq \exp\exp\exp(2H^{10^{n^{10}}}).$$

Baker's method applies to other types of equations, for instance to exponential equations. The most striking example is the result of Tijdeman on the Catalan equation: every solution $(x,y,z,t) \in \mathbf{Z}^4$ in integers ≥ 2 of the equation $x^y - z^t = 1$ can be effectively bounded:

$$x^y \leq \exp\exp\exp 250.$$

See [Ti], [Lg].

8.5. Applications to elliptic curves with good reduction outside a given finite set of places.

For many modular curves, one can construct non-constant morphisms to $\mathbf{P}_1 - \{0,1,\infty\}$ (see Exercise below).

The first example is that of the curve $X(2)$ which classifies elliptic curves equipped with a basis of points of order 2. The parameter which is classically called λ gives a morphism $\lambda : \bar{X}(2) \to \mathbf{P}_1$ where $\bar{X}(2)$ is the completed modular curve i.e. with cusps added. Writing an elliptic curve in Legendre form

$$y^2 = x(x-1)(x-\lambda),$$

λ is connected with the j-invariant by

$$j = 2^8 \frac{(\lambda^2 - \lambda + 1)^3}{\lambda^2(\lambda-1)^2}.$$

The cusps are $\lambda = 0, 1, \infty$ so that $X(2) = \mathbf{P}_1 - \{0,1,\infty\}$. In particular, every quasi-integral set of points of $X(2)$ is finite.

Here is an application. Recall that an elliptic curve has good reduction at a place v if there is a model of the curve

$$y^2 + a_1xy + a_3y = x^3 + a_2x^2 + a_4x + a_6,$$

where the a_i are v-integers, and the discriminant is invertible at v. Reduction modulo v then gives an elliptic curve.

Theorem. *Let K be a number field and S a finite set of places of K. Then up to isomorphism, there are only a finite number of elliptic curves over K with good reduction outside S and these can be found effectively.*

First proof.

We can assume that S contains the prime divisors of 2. Let E be an elliptic curve having good reduction outside S. Let K' be the field obtained by adjoining to K the points of order 2 of E (if we write the equation of E as $y^2 = f(x)$, K' is obtained by adjoining the roots of f to K). Then the extension K'/K is unramified outside S. By Hermite's theorem and by making a finite extension of K, we may assume that the points of order 2 are in the field K. In other words, the curve can be written in the Legendre form

$$y^2 = x(x-1)(x-\lambda).$$

The set of λ so obtained is quasi-integral in $X(2)$, for j is integral in A_S and the inverse image of ∞ by the morphism $\mathbf{P}_1 \to \mathbf{P}_1$, $\lambda \mapsto j$ is $\{0,1,\infty\}$.

Thus the set of these λ is finite and can be found effectively by Baker's method (cf. §8.3).

Second proof.

We may enlarge S so that it contains the divisors of 2 and 3, and that A_S is principal. One checks that the equation of the curve can then be written as

$$y^2 = x^3 + Ax + B,$$

where the discriminant $\Delta = -4A^3 - 27B^2$ belongs to A_S^*. As Δ is defined modulo the 12th powers, this leads to a finite number of equations $4x^3 + 27y^2 = \text{const}$ (see [Se3, IV.1.4] for more details), to which one applies Case 2 of §8.4.

Problem. What are the elliptic curves over $\mathbf{Q}$ with good reduction outside 11? (Those with conductor 11 are known, but not those with conductor 11^2, as far as I know.)

Corollary. *Let E be an elliptic curve defined over K. There are only finitely many elliptic curves E' over K, up to isomorphism, which are K-isogenous to E.*

To see this, we take S so large that E has good reduction outside S. By a general property of abelian varieties, the same holds for E'. From the previous theorem, the E' are finite in number.

It would be interesting to have an explicit bound for this number (for $K = \mathbf{Q}$, M. Kenku [Ke1] has shown that 8 is the precise bound).

Remark. The above corollary can be used (see [Se3, Ch.IV, §2.1]) to give another proof of the irreducibility theorem (§5.4) that if E does not have complex multiplication then the action of $\mathrm{Gal}(\bar{K}/K)$ on the Tate module $V_l(E)$ is irreducible.

Exercise. (Kubert-Lang). Let $X_1(n)$ be the modular curve which classifies pairs (E, P) where E is an elliptic curve and P a point of order n on E. Assume $n \geq 6$. Construct a non-constant morphism

$$\phi : X_1(n) \to \mathbf{P}_1,$$

by

$$\phi(E, P) = \frac{x(P) - x(2P)}{x(P) - x(3P)},$$

where x is the "x-coordinate" on E. Show that ϕ takes values in $\mathbf{P}_1 - \{0, 1, \infty\}$ and deduce from that the effective finiteness of quasi-integral sets on $X_1(n)$.

9. HILBERT'S IRREDUCIBILITY THEOREM

Hilbert's irreducibility theorem has quite a large number of proofs, based on different principles; some give precise estimates for the number of integers that one seeks. There are several applications: the construction of elliptic curves over $\mathbf{Q}$ having rank ≥ 9 and the construction of extensions of $\mathbf{Q}$ with Galois groups S_n, A_n,...

9.1. Thin sets.

Let k be a field of characteristic 0. Let Ω be a subset of $\mathbf{P}_n(k)$ or of $\mathbf{A}_n(k) = k^n$.

Definition. *We say that Ω is "thin" (Fr.: "ensemble mince") if there is an algebraic variety X over k and a morphism $\pi : X \to \mathbf{P}_n$ (resp. $\pi : X \to \mathbf{A}_n$) with the following properties:*
(a) $\Omega \subset \pi(X(k))$,
(b) The fibre of π over the generic point is finite and π has no rational section over k (i.e. in a neighbourhood of the generic point π is quasi-finite and has no section).

A subset of $\mathbf{P}_n(k)$ is thin if and only if its intersection with $\mathbf{A}_n$ is thin. Any finite union of thin sets is thin.

There are two important types of thin sets. We shall say that a subset Ω has *type 1* if it is contained in a Zariski-closed subvariety F, different from the entire space. To see that such a subset Ω is thin, take $X = F$ and π the injection. The generic fibre is empty.

We shall call a subset *type 2* if it has the form $\pi(X(k))$, where X is an irreducible variety over k of dimension n and π is a dominating morphism $X \to \mathbf{P}_n$ or $X \to \mathbf{A}_n$ of degree $d \geq 2$. This is the most interesting case of a thin set. If we take $t_1, ..., t_n$ as affine coordinates, the function field $k(X)$ of X is an extension of $k(t_1, ..., t_n)$ of degree d. If X is not absolutely irreducible, then $\pi(X(k))$ is of type 1. Indeed, in that case the field $k(X)$ contains a subfield k_1 finite over k and $\neq k$. On the other hand, if $x \in X(k)$, the local ring of x does not contain k_1, hence is not normal; this shows that $X(k)$ is contained in the non-normal locus of X, and $\pi(X(k))$ is of type 1. [An example is the variety $x^2 + y^2 = 0$ over $\mathbf{R}$, which defines an irreducible, but not absolutely irreducible, variety; $\mathbf{R}(i) = \mathbf{C}$ is in the function field. There is only one point over $\mathbf{R}$; the normalised variety has no point over $\mathbf{R}$.]

If necessary, we may then assume that X is absolutely irreducible. The following statement is obvious.

Proposition. *Every thin subset of $\mathbf{P}_n(k)$ or $\mathbf{A}_n(k)$ is contained in a finite union of subsets of types 1 and 2.*

Examples of thin sets.

Let $d \geq 2$ be an integer. In $\mathbf{A}_1(k) = k$, the subset $k^d = \{x^d, x \in k\}$ is a thin subset (take $X = \mathbf{A}_1$, and $\pi : X \to \mathbf{A}_1$ defined by $\pi(x) = x^d$).

A finite union $\bigcup c_i k^d$ of classes modulo k^d is also a thin set.

Polynomial interpretation.

The definitions above can be reformulated as follows for $\mathbf{A}_n$, say.

That Ω is thin of type 1 is equivalent to the existence of a non-zero polynomial $G(T_1, ..., T_n)$, with coefficients in k, such that $G(t_1, ..., t_n) = 0$ for all $(t_1, ..., t_n) \in \Omega$.

For type 2, we take a polynomial $F(X, T) \in k[T]$ where $T = (T_1, ..., T_n)$. We assume that F is absolutely irreducible, that is to say, irreducible in $\bar{k}(T)[X]$ (where $\bar{k}$ is an algebraic closure of k). We also assume that F has degree ≥ 2 in X. Then the set

$$\Omega_F = \left\{ t; \quad \begin{array}{c} t \text{ is not a pole of the coefficients of } F, \\ \\ \text{and} \quad F(X,t) \text{ has a root in } k \end{array} \right\}$$

is thin of type 2, and we obtain all of them thus (modulo type 1). [If we take an irreducible variety as in the definition of type 2, its function field is a finite separable extension of $k(T)$, and we choose for F a generator of that extension.]

9.2. Specialisation of Galois groups.

We shall see that specialisation outside a thin set preserves the irreducibility of a polynomial and the structure of its Galois group.

For the first statement, we assume that X is irreducible over k, of dimension n, and $\pi : X \to \mathbf{P}_n$ is dominating.

Proposition 1. *There is a thin set Ω in $\mathbf{P}_n(k)$ such that for all $t \notin \Omega$, $\pi^{-1}(t)$ consists of a finite number of points which are conjugate over k.*

In terms of polynomials, this means that if $F(X; T_1, ..., T_n)$ is an irreducible polynomial over $k(T)$, then for all $t = (t_1, ..., t_n) \notin \Omega$, $F(X, t)$ is irreducible over k. More precisely, it follows from Proposition 2 below that *$F(X, t)$ has the same Galois group as $F(X, T)$ outside a thin set.*

For the second statement, start again with $\pi : X \to \mathbf{P}_n$ with X irreducible, and consider a finite group G acting faithfully on X such that X/G is identified by the projection π with an open subscheme U of $\mathbf{P}_n$. We take U small enough so that π is étale. Let u be a rational point of U and select a closed point x of the scheme X above u. Up to conjugation, the subgroup G_x of G which stabilises x, depends only on u. The extension $k(x)/k$ is Galois with Galois group G_x.

Proposition 2. *There is a thin set Ω of $\mathbf{P}_n(k)$ outside of which $G_x = G$.*

Proof of Proposition 2.
Let $\{H_i\}$ be the subgroups of G different from G (one can restrict oneself to maximal subgroups). Define $X_i = X/H_i$; this is a variety which projects onto X/G, thus generically on $\mathbf{P}_n$, and the degree of the projection is $[G : H_i] > 1$. The union Ω of the images of the $(X/H_i)(k)$ is a thin set. Proposition 2 follows.

Proof of Proposition 1.
It suffices to embed the field extension corresponding to $X \to \mathbf{P}_n$ in a Galois extension, and to use proposition 2.

9.3. Examples of degrees 2,3,4,5.

We consider a polynomial $F(X; T_1, ..., T_n) = F(X; T)$ irreducible over $k(T)$, of degree $d \geq 2$ in X and we shall describe a thin set $\Omega \subset k^n$ such that for $t \notin \Omega$, $F(X; t)$ has the same Galois group as $F(X; T)$.
1). $d = 2$. One reduces to the equation $X^2 = g(T)$, where $g(T) \in k(T)$ is not a square in $k(T)$. We take

$$\Omega = \{t; g(t) \text{ is a square in } k\}.$$

2). $d = 3$, the Galois group of $F(X; T)$ being S_3. The proper maximal subgroups of S_3 are A_3 and the subgroups of order 2. The group A_3 has index 2 in S_3, the corresponding extension is given by the square root of the discriminant $\Delta(T)$ of $F(X; T)$. The first condition (relative to A_3) is that $\Delta(t)$ is not a square in k (note that by assumption $\Delta(T)$ is not a square in $k(T)$). The second condition (relative to the subgroups of order 2) is that $F(X; t)$ has no root in k (that is to say, is irreducible over k since $d = 3$). Hence, we take

$$\Omega = \{t; \Delta(t) \text{ is a square in } k\} \cup \{t; F(X; t) \text{ has a root in } k\}.$$

3). $d = 4$, the Galois group of $F(X;T)$ being S_4. The proper maximal subgroups are: the alternating group A_4, the groups S_3 fixing one letter, and the Sylow 2-subgroups D_4 dihedral of order 8.

The condition relative to A_4 is that the discriminant is not a square, that relative to S_3 is that the equation has no root in k.

The subgroup D_4 appears as the stabiliser of the decomposition of a set of 4 elements in 2 pairs:

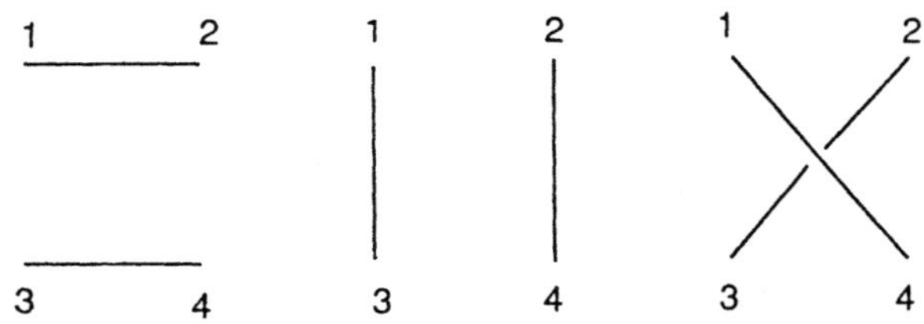

We introduce the expressions

$$x_1x_2 + x_3x_4, x_1x_3 + x_2x_4, x_1x_4 + x_2x_3,$$

and write the unique monic equation having these 3 roots (the cubic resolvent). The condition relative to D_4 is that the cubic resolvent has no root in k.

The cubic resolvent has the same discriminant as the initial equation, because

$$x_1x_2 + x_3x_4 - x_1x_3 - x_2x_4 = (x_1 - x_4)(x_2 - x_3).$$

Therefore if the roots x_i of F are distinct, the roots of the cubic resolvent are distinct. (We then see that the above thin set is the smallest appropriate.)

4). $d = 5$, the Galois group of $F(X;T)$ being S_5.

The maximal subgroups of S_5 are A_5, S_4 (the stabiliser of one letter), $S_2 \times S_3$, and H_{20} the normaliser of a Sylow 5-subgroup, of index 6.

The conditions on A_5 and S_4 require that F and the equation $t^2 = \Delta$ have no root in k.

The condition relative to $S_2 \times S_3$ means that the polynomial $F(X,t)$ does not decompose as a product of a quadratic factor and a cubic factor. If $x_1, ..., x_5$ are the roots (assumed distinct, since $\Delta \neq 0$), we write the equation of degree 10 whose roots are the $x_i + x_j$, $(i \neq j)$. We then take t outside the thin set for which this equation has a root.

The group H_{20} can be described as

$$\{x \mapsto ax + b; b \in \mathbf{Z}/5\mathbf{Z}, a \in (\mathbf{Z}/5\mathbf{Z})^*\},$$

if S_5 is realised as the group of permutations of $\mathbf{Z}/5\mathbf{Z}$. A Sylow 5-subgroup of S_5 corresponds to a pentagonal arrangement

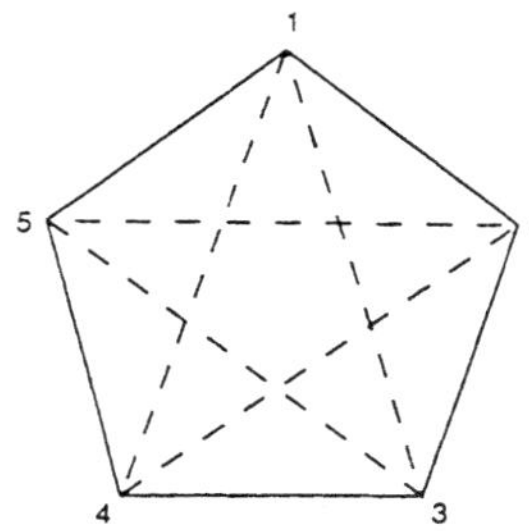

Given such an arrangement of $\{1,2,3,4,5\}$ we introduce the following expression:

$$u = x_1x_2 + x_2x_3 + x_3x_4 + x_4x_5 + x_5x_1 - \{x_1x_3 + x_3x_5 + x_5x_2 + x_2x_4 + x_4x_1\},$$

(the signs + correspond to the sides of the convex polygon, the signs - correspond to the sides of the pentagonal star). As this expression is anti-invariant under the permutations of H_{20} we divide it by

$$d = \prod_{i<j}(x_i - x_j),$$

and write the equation of degree 6 (the sextic resolvent) whose 6 roots are $\alpha = u/d$ and the 5 other analogous expressions.

Theorem. *Assume that $\Delta \neq 0$ and that the Galois group G of $F(X,t)$ is not contained in one of the groups S_4, $S_2 \times S_3$. Then G is contained in some H_{20} (i.e. a normaliser of a Sylow 5-subgroup) if and only if one of the α is in the ground field.*

Therefore to check the condition relative to H_{20}, it is necessary and sufficient to check that the sextic resolvent has no rational root.

The hypothesis $G \not\subset S_4$ and $G \not\subset S_2 \times S_3$ (up to conjugation) means that G is transitive, i.e. that the polynomial is irreducible.

Proof of theorem.

If G is contained in some H_{20}, then the corresponding α is in the ground field.

Conversely, suppose that one of the α is in k and that G is not contained in any H_{20}. As G is transitive, its order is divisible by 5, and the number of its Sylow 5-subgroups is $\equiv 1 \bmod 5$. Therefore all the Sylow 5-subgroups of S_5 (there are 6 of them) are contained in G; hence G contains A_5. But A_5 permutes transitively the α's; the α's are thus conjugate over k. By hypothesis one of them is in k. Therefore

$$\alpha_1 = \alpha_2 = \alpha_3 = \alpha_4 = \alpha_5 = \alpha_6.$$

We have

$$\alpha_1 = \frac{1}{d}\{x_1x_2 + x_2x_3 + x_3x_4 + x_4x_5 + x_5x_1 - (x_1x_3 + x_3x_5 + x_5x_2 + x_2x_4 + x_4x_1)\}.$$

We permute 1 and 2 and use the fact that d is anti-invariant. This gives

$$\alpha_2 = \frac{1}{d}\{x_1x_4 + x_1x_5 + x_2x_4 + x_2x_3 + x_3x_5 - (x_1x_2 + x_1x_3 + x_3x_4 + x_4x_5 + x_5x_2)\}.$$

The equation $\frac{d}{2}(\alpha_1 - \alpha_2) = 0$ becomes

$$x_1x_2 + x_3x_4 + x_4x_5 - x_1x_4 - x_2x_4 - x_3x_5 = 0,$$

that is to say

$$x_1x_2 + x_3x_4 + x_4x_5 = x_1x_4 + x_2x_4 + x_3x_5.$$

Similarly, we have:

$$x_1x_3 + x_2x_4 + x_4x_5 = x_2x_5 + x_4x_1 + x_4x_3.$$

Subtracting the preceding equation, we get

$$x_1x_3 + x_2x_4 - x_1x_2 - x_3x_4 = x_2x_5 + x_4x_3 - x_2x_4 - x_3x_5,$$

i.e.

$$(2x_4 - x_1 - x_5)(x_2 - x_3) = 0.$$

As the x_i are distinct this implies

$$2x_4 = x_1 + x_5.$$

But by permuting the variables we also have

$$2x_3 = x_1 + x_5,$$

therefore $x_4 = x_3$, which contradicts $\Delta \neq 0$.

9.4. Further properties of thin sets.

Plane sections.

Theorem. *Let n and d be integers, $n \geq d \geq 1$, and let Ω be a thin subset of $\mathbf{P}_n(k)$. Then there is a non-empty open subscheme U of the Grassmannian $G_{n,d}$ of d-dimensional linear subspaces of $\mathbf{P}_n$, such that $\Omega \cap L$ is thin for all $L \in U(k)$.*

The most interesting case is $d = 1$. For example, if Ω is a thin subset of $\mathbf{P}_2$, then the intersection of Ω with a line is almost always thin.

This theorem can be used to reduce many questions on thin sets to dimension 1.

Proof.

We may assume Ω is of type 1 or 2, and type 1 poses no problem.

Let then X be absolutely irreducible with a projection onto $\mathbf{P}_n$ of degree ≥ 2; let L be a linear variety of dimension d and let X_L be the pullback over L:

$$\begin{array}{ccc} X_L & \subset & X \\ \downarrow & & \downarrow \\ L & \subset & \mathbf{P}_n \end{array}$$

If X_L is irreducible, as the degree of $X_L \to L$ is equal to the degree of $X \to \mathbf{P}_n$ we see that $\Omega \cap L$ is thin.

It remains to tell for which linear varieties $L \in G_{n,d}$ the variety X_L is irreducible. The answer is provided by Bertini's theorem: this is true for a non-empty open subset of $G_{n,d}$ (see [H, Chapter II, Theorem 8.18]).

Remark. The linear subspaces L for which $\Omega \cap L$ is thin can be described more precisely. As the given covering of $\mathbf{P}_n$ is non-trivial, it ramifies over a non-empty hypersurface H. If L is a line which only intersects H at smooth points of H, and transversally at each of these points, then $X_L \to L$ is irreducible (thus $\Omega \cap L$ is thin).

Over $\mathbf{C}$, we can see this by topological arguments: the property is true for the generic line and if a line L of the preceding type is moved slightly, one passes from L to its transform by a homeomorphism.

Example. $X =$ "Double plane": $T^2 = \phi$, where ϕ is a polynomial of degree 4 defining a non-singular quartic of $\mathbf{P}_2$.

The covering is ramified along the quartic. If L cuts the quartic in 4 distinct points, the intersection has for its equation $T^2 =$ polynomial with 4 simple roots, that is to say an elliptic curve.

If L is an ordinary tangent cutting the quartic in 2 other distinct points, or if L is an inflection tangent, we obtain an equation T^2 = polynomial of degree 2, thus a curve of genus 0, which projects on the line with degree 2.

Finally, if L is one of the 28 bitangents, we obtain an equation T^2 = a square, which decomposes in two curves of genus 0.

We can therefore take U to be all the lines minus the 28 bitangents.

Extension of the ground field.

Proposition. *Let k_1 be a finite extension of k. If $\Omega_1 \subset \mathbf{P}_n(k_1)$ is thin over k_1, then $\Omega = \Omega_1 \cap \mathbf{P}_n(k)$ is thin over k.*

Proof.

It is enough to prove this with $\mathbf{P}_n$ replaced by $\mathbf{A}_n$. We may assume that $\Omega_1 = \pi_1(X_1(k_1))$, where $\pi_1 : X_1 \to \mathbf{A}_n$ is a k-morphism without a rational section, X_1 being some k-variety, which we may suppose to be affine. This last assumption allows us to apply the "restriction of scalars" functor $R_{k_1/k}$, which transforms it to an affine k-variety $X = R_{k_1/k}X_1$.

[A characteristic property of X is that it represents the functor $Y \mapsto \mathrm{Morph}_{k_1}(Y, X_1)$, i.e. for every k-variety Y, we have a natural bijection $\mathrm{Morph}_k(Y, X) = \mathrm{Morph}_{k_1}(Y, X_1)$. Taking Y to be a point, this gives $X(k) = X_1(k_1)$. For more details, see [Oe], App. 2, where $R_{k_1/k}$ is denoted $\prod_{k_1/k}$ and is defined in a more general setting.]

By applying this functor to the k_1-morphism $\pi_1 : X_1 \to \mathbf{A}_n$ we get a k-morphism $\pi : X \to Y$, where $Y = R_{k_1/k}(\mathbf{A}_n)$ (which is isomorphic to $\mathbf{A}_{dn}$, if $d = [k_1 : k]$). But since $\mathbf{A}_n$ is a k-variety, there is a "diagonal" morphism $\beta : \mathbf{A}_n \to Y$ ([Oe], A.2.6), which on k-points corresponds to the natural injection

$$\mathbf{A}_n(k) \to Y(k) = R_{k_1/k}X_1.$$

Since $\Omega = \mathbf{A}_n(k) \cap \Omega_1$, we have $\Omega = \beta^{-1}(\Omega_1) = \beta^{-1}(\pi(X(k))$. Let us now define Z by the Cartesian diagram

$$\begin{array}{ccc} Z & \to & X = R_{k_1/k}X_1 \\ \downarrow & & \downarrow \\ \mathbf{A}_n & \to & Y = R_{k_1/k}\mathbf{A}_n \end{array}$$

If ϕ is the projection of Z on $\mathbf{A}_n$, we have $\Omega = \phi(Z(k))$. Hence to prove that Ω is thin, it is enough to check that the k-morphism $\phi : Z \to \mathbf{A}_n$ does not have any rational section. But if $s : \mathbf{A}_n \to Z$ were such a section, it would give a rational map $\alpha \circ s : \mathbf{A}_n \to X$, which, by the universal property of X (extended to rational maps) would give a k_1-rational map $s_1 : \mathbf{A}_n \to X_1$

which would be a rational section of π_1, contrary to the hypotheses. This concludes the proof.

Example. Take the fields to be $\mathbf{Q}$ and $\mathbf{Q}(i)$ and the equation $z^2 = it_1$ defining a thin set Ω_1 over $\mathbf{Q}(i)$; we seek the $t_1 \in \mathbf{Q}$ such that there is a solution:

$$(x+iy)^2 = it_1,$$

that is to say

$$x^2 - y^2 = 0, 2xy = t_1.$$

We eliminate y:

$$4x^4 - t_1^2 = 0.$$

This is an equation with rational coefficients, which factorises as

$$(2x^2 - t_1)(2x^2 + t_1) = 0.$$

As $2i$ is a square in $\mathbf{Q}(i)$, these factors over $\mathbf{Q}(i)$ are simply $z^2 - it_1$ and $z^2 + it_1$.

Exercise. Let X be an irreducible algebraic variety over k, and $\pi : X \to \mathbf{P}_n$ a dominating morphism. Show that there is a thin set in $\mathbf{P}_n$ outside of which the fibres are irreducible (over k).

(First take the case where X has dimension n, then for the general case, factorise π birationally in order to have first an absolutely irreducible fibre and then a finite fibre.)

9.5. Hilbertian fields.

Definition. *A field k is called hilbertian if for all $n \geq 1$, $\mathbf{P}_n(k)$ is not thin.*

Remarks.

1). For k to be hilbertian, it is enough that $\mathbf{P}_1(k)$ is not thin. This follows from the theorem of §9.4. One might also replace $\mathbf{P}_1(k)$ by $\mathbf{A}_1(k)$.

2). If k is hilbertian, *the complement of a thin set Ω of $\mathbf{P}_n(k)$ is Zariski dense* (otherwise $\mathbf{P}_n(k)$ would be the union of Ω and a proper closed subset, hence would be thin).

3). If k is hilbertian, then for every integer $d \geq 2$, k^*/k^{*d} is infinite.

In particular a local field is not hilbertian (neither is an algebraically closed field). Thus $\mathbf{Q}_p$, $\mathbf{R}$, $\mathbf{C}$ are not hilbertian.

4). If k is hilbertian and k_1 is finite over k, then k_1 is hilbertian (this follows from the last proposition of §9.4).

The converse does not always hold ([FJ], Ex. 12.17).
5). For every field K, the field $K(T)$ is hilbertian. Indeed, let Ω be a thin set in $K(T)$. Then the intersection of Ω with the set of polynomials $aT+b$ of degree 1 with coefficients in K is contained in a closed subset $\neq K \times K$ for the Zariski topology on $K \times K$ (this result follows directly from Bertini's theorem (see [Ro])). Hence Ω cannot be equal to $K(T)$.

Therefore, if k is hilbertian, every finitely generated extension of k is hilbertian.

For more examples of hilbertian fields see [FJ, Chapter12].

9.6. The irreducibility theorem: elementary proof.

Hilbert's Theorem. *The field* $\mathbf{Q}$ *is hilbertian.*

Hence number fields, and more generally finitely generated extensions of $\mathbf{Q}$, are hilbertian.
Proof.

Here is a variant of Lang's proof [L2, Ch. VIII] which is itself a variant of Hilbert's proof. We consider a thin set in the affine line $\mathbf{A}_1(\mathbf{Q}) = \mathbf{Q}$, and we examine its intersection with $\mathbf{Z}$, say Ω. For $B > 0$, let ω_B be the number of $\omega \in \Omega$ with $|\omega| \leq B$. We prove the theorem in the following refined form.

Theorem. *We have*

$$\omega_B = O(B^{1-\epsilon}), \text{ for some } \epsilon > 0.$$

The proof will give ϵ and O effectively. (We shall see in §9.7 that ϵ can be taken to be $\frac{1}{2}$.)

We may assume that we have a curve X which is absolutely irreducible, and a finite morphism $X \to \mathbf{A}_1$ of degree ≥ 2, and that the points of Ω lift to rational points of X. We may complete X to a projective curve $\bar{X}$ above $\mathbf{P}_1$, and we write $X_\infty = \bar{X} - X$. The points of Ω lift to rational points of X and form a quasi-integral set. In the affine ring of X, take a function ϕ which is not in the affine ring of $\mathbf{A}_1$ (that is to say, which is not a polynomial). We take a branch of the curve X in a neighbourhood of infinity and take the corresponding Puiseux expansion of ϕ:

$$\phi(x) = \sum_{\lambda} c_\lambda x^\lambda, \lambda \to -\infty, c_\lambda \in \mathbf{R};$$

the series converges for sufficiently large real x.

Multiplying ϕ by a constant, we may assume that $\phi(x) \in \mathbf{Z}$ for all $x \in \Omega$. (This follows from the fact that ϕ is integral over $\mathbf{Q}[X]$.) We are thus led to prove the following statement.

Theorem. *Let*

$$\phi(t) = \sum_{\lambda \to -\infty} c_\lambda t^\lambda, (c_\lambda \in \mathbf{R}, \lambda \in \mathbf{Q})$$

be a Puiseux expansion in the neighbourhood of infinity, convergent for $t \geq R$. *Let* Ω_ϕ *be the set of* $t \geq R$ *such that* $t \in \mathbf{Z}$ *and* $\phi(t) \in \mathbf{Z}$. *If* ϕ *is not a polynomial, then the number of* $t \in \Omega_\phi$ *with* $|t| \leq B$ *is* $O(B^{1-\epsilon})$ *for some* $\epsilon > 0$.

Proof.

There is an integer $n \geq 1$ such that all the exponents of the series for the $n-1$'st derivative $\phi^{(n-1)}(t)$ are < 0, and this series is not identically zero. In particular

$$\phi^{(n-1)}(t) \sim c_1 t^{-\mu}, \text{ as } t \to \infty, \text{ with } \mu > 0, c_1 \neq 0.$$

The idea of the proof is to show that n points of Ω_ϕ cannot be too close; more precisely

Lemma. *There are two constants* $\alpha > 0$, $c > 0$ *such that if* t *is sufficiently large, the interval* $[t, t + ct^\alpha]$ *contains at most* $n-1$ *points of* Ω_ϕ.

Let $t_1 < \ldots < t_n$ be sufficiently large elements in Ω_ϕ, and $y_i = \phi(t_i)$. We construct the interpolation polynomial P of degree $n-1$ such that $P(t_i) = y_i$; this is given explicitly by the Lagrange formula

$$P(T) = \sum_{j=1}^{n} y_j \frac{\prod_{1 \leq i \leq n, i \neq j}(T - t_i)}{\prod_{1 \leq i \leq n, i \neq j}(t_j - t_i)}.$$

Then $\phi - P$ vanishes on $t_1, \ldots, t_n$. By Rolle's theorem, there is $\xi \in [t_1, t_n]$ such that

$$\phi^{(n-1)}(\xi) = P^{(n-1)}(\xi).$$

As P has degree $n-1$, $P^{(n-1)}(\xi)$ is the highest degree term multiplied by $(n-1)!$

$$P^{(n-1)}(\xi) = (n-1)! \sum_{j=1}^{n} \frac{y_j}{\prod_{1 \leq i \leq n, i \neq j}(t_i - t_j)}.$$

This is a rational number whose denominator is at most

$$|\prod_{1 \leq i < j \leq n} (t_i - t_j)| < (t_n - t_1)^{n(n-1)/2}.$$

On the other hand, as t_1 is sufficiently large, we have

$$0 < |\phi^{(n-1)}(\xi)| \le c_2 t_1^{-\mu}.$$

If we put $l = t_n - t_1$, we have $|P^{(n-1)}(\xi)| \ge 1/l^{n(n-1)/2}$ i.e.

$$1 \le l^{n(n-1)/2} |\phi^{(n-1)}(\xi)|,$$

whence

$$l \ge c t_1^{\alpha}, \text{ with } \alpha = 2\mu/n(n-1), \text{ and } c = c_2^{-2/n(n-1)}.$$

We deduce from this lemma an upper bound for the number of points of Ω_ϕ between 1 and B in the following way. We put $\epsilon = 1/(1+\alpha) > 0$. Cut $[1, B]$ into $[1, B^\epsilon]$ and $[B^\epsilon, B]$, then decompose $[B^\epsilon, B]$ into $O(B/B^{\alpha\epsilon})$ equal intervals of length $< cB^{\alpha\epsilon}$; in each of these small intervals there are at most $O(1)$ points of Ω_ϕ, thus in $[1, B]$ there are at most $O(B^\epsilon + B^{1-\alpha\epsilon})$ points of Ω_ϕ, which gives $O(B^\epsilon)$ since $\epsilon = 1/(1+\alpha)$. As $\alpha > 0$, we have $\epsilon < 1$.

The estimate $O(B^{1-\epsilon})$ for ω_B shows that the complement of a thin set of $\mathbf{Z}$ contains infinitely many prime numbers, infinitely many sums of two squares etc.

9.7. Thin sets in $\mathbf{P}_1$: upper bounds.

We continue the study of thin sets by showing that over a number field these sets are "small". We consider only the case of 1 variable; for the general case see chapter 13.

Let K be a number field of degree d over $\mathbf{Q}$. Let X be an irreducible algebraic curve over K. Consider a morphism $\pi : X \to \mathbf{P}_1$ of finite degree $\delta \ge 2$. We wish to show that $\pi(X(K))$ is small in a suitable sense in $\mathbf{P}_1(K) = K \cup \{\infty\}$. We do this by two different methods.

First method.

We count the rational points of bounded height. For $B \ge 2$, denote by $N(B)$ the number of $t \in \mathbf{P}_1(K)$ with $H(t) \le B$, and $N_X(B)$ the number of $t \in \pi(X(K))$ with $H(t) \le B$.

With our normalisation of the height, Schanuel's theorem (§2.5) gives us: $N(B) \sim cB^{2d}$, for some $c > 0$. For example, over $\mathbf{Q}$ we have $d = 1$, $c = 12/\pi^2$.

Theorem. *We have* $N_X(B) = O(B^d)$ *as* $B \to \infty$.

Proof.

This is a simple exercise on heights. We may assume that the curve X is smooth and projective (this only changes a finite number of points). We then have a covering map $\pi : X \to \mathbf{P}_1$ which gives a height on X: $H(x) = H_{\mathbf{P}_1}(\pi(x))$. This is the height on X associated to $\pi^* D$ where D is the ample line bundle $\mathbf{O}(1)$ on $\mathbf{P}_1$. Thus up to a bounded factor, $N_X(B)$ is the number of $x \in X(K)$ with $H(x) \leq B$.

There are essentially 3 cases to consider.

Case 0: X has genus 0.

If X has no rational point, we have $N_X(B) = 0$. If $X(K) \neq \emptyset$ then X is a projective line. Let $\phi : X \to \mathbf{P}_1$ be an isomorphism. The divisor $\pi^* D$ has degree equal to the degree δ of π and the associated height is equivalent to the δ-th power of the height on $\mathbf{P}_1$:

$$H(x) \asymp H(\phi(x))^\delta, \text{ for } x \in \mathbf{P}_1(K).$$

Therefore

$$N_X(B) \asymp (B^{1/\delta})^{2d} = O(B^d),$$

since $\delta \geq 2$.

Case 1: X has genus 1.

If $X(K) \neq \emptyset$, we take a point of $X(K)$ as an origin, and we have an elliptic curve. Let ρ be the rank of $X(K)$. Then

$$N_X(B) \asymp (\log B)^{\rho/2},$$

(cf.§4.5).

Case 2: X has genus ≥ 2.

We have seen by Mumford's theorem (§5.7) that

$$N_X(B) = O(\log \log B).$$

[Mordell's conjecture gives $N_X(B) = O(1)$.]

The theorem is therefore proved.

Note that only in the case of genus 0 does $N_X(B)$ increase as a power of B and in this case the result is effective.

It ought not to be difficult to prove an effective upper bound of the form $O((\log B)^A)$ in the case of genus ≥ 1. For elliptic curves, we do not know how to compute ρ effectively, but we know how to bound it. One uses then a lower bound for $\tilde{h}(x)$ when $\tilde{h}$ is the Néron normalised height and x is a non-torsion rational point (cf. [Ma2]). For genus ≥ 2, one can bound $N_X(B)$ just by working in the jacobian.

Second method.

We count the number of integral points, as in Hilbert's proof (we would also be able to count the S-integral points).

Let A be the ring of integers of K. For $\alpha \in A$, we write $||\alpha|| = \sup |\sigma\alpha|$ where σ runs over the the embeddings of K in $\mathbf{C}$. An equivalent norm is obtained by writing $A \cong \mathbf{Z}^d$ and imposing a norm on $\mathbf{R}^d$. The number $N(B)$ of $\alpha \in A$ with $||\alpha|| \leq B$ satisfies

$$N(B) \asymp B^d.$$

Let Ω be a thin set, $\Omega \subset A$. We define

$$N_\Omega(B) = \text{ number of } \alpha \in \Omega \text{ with } ||\alpha|| \leq B.$$

Theorem. *We have* $N_\Omega(B) = O(B^{d/2})$.

For example for $K = \mathbf{Q}$, if Ω is a thin set contained in $\mathbf{Z}$, the number of points of Ω between $-B$ and $+B$ is $O(B^{1/2})$. This estimate is best possible, since $\{n^2, n \in \mathbf{Z}\}$ is a thin set. The estimate obtained above in §9.6 was $O(B^{1-\epsilon})$ for some $\epsilon > 0$.

Proof.

The principle is the same as for rational points. Let X_∞ be the inverse image of $\infty \in \mathbf{P}_1$. The affine curve $X_{\text{aff}} = X - X_\infty$ projects onto $\mathbf{A}_1 = \mathbf{P}_1 - \{\infty\}$ by a finite, and thus proper, morphism.

Let Ω_X be the set of rational points of X whose image by π is an integral point of $\mathbf{A}_1$ and let Ω be the image of Ω_X. Since π is proper, Ω_X is a quasi-integral set in X_{aff}. By Siegel's theorem, Ω_X is finite (thus Ω is also) except when the genus of X is zero, and X_{aff} has just 1 or 2 points. Let us consider these two exceptional cases.

First case: X has genus 0 and $X_\infty = \{P\}$.

As X_∞ reduces to a point, this point is rational. We can therefore identify X_{aff} with $\mathbf{A}_1$. Then π is given by a polynomial: $\pi(x) = \alpha_0 x^m + \ldots + \alpha_m$, with $\alpha_i \in K$, and $\alpha_0 \neq 0$. We have $m = \delta$ (the degree of the polynomial is equal to the degree of the morphism). Let c be a positive integer such that the $c\alpha_i$ are integers, $(0 \le i \le m)$. If $\pi(x)$ is an integer with $x \in K$, then $c\alpha_0 x$ is an integer. To count the $x \in K$ such that $\pi(x) \in A$ and $||\pi(x)|| \le B$, it suffices to count the $x \in A$ such that $||x|| \le rB^{1/m}$, for some fixed $r > 0$: their number is $\asymp B^{d/m}$. Therefore

$$N_\Omega(B) = O(B^{d/m}), \text{ with } m \ge 2.$$

Second case: X has genus 0 and 2 points at infinity.

After a ground field extension, X_{aff} becomes isomorphic to $\mathbf{P}_1 - \{0, \infty\}$. The set U of units, corresponding to the affine equation $xy = 1$, is an integral set, and the quasi-integral sets are finite unions $\bigcup_i c_i U$. We are thus led to count the units of bounded height. We take fundamental units $\epsilon_1, ..., \epsilon_\lambda$, and write $\epsilon = w\epsilon_1^{n_1}...\epsilon_\lambda^{n_\lambda}$, where w is a root of unity, and $n_i \in \mathbf{Z}$. For $c \in K$, $c \ne 0$ fixed, we have $\log H(c\epsilon) \asymp \sup |n_i|$. Therefore the number of units ϵ with $H(c\epsilon) \le M$ is $O((\log M)^\lambda)$.

Under the morphism $X \cong \mathbf{P}_1 \to \mathbf{P}_1$, any height on $X \cong \mathbf{P}_1$ is a power of the standard height on $\mathbf{P}_1$:

$$\log H(x) \asymp \log H(\pi(x)) \asymp \log ||\pi(x)||.$$

We take M to be a power of B. Then

$$N_\Omega(B) = O((\log B)^\lambda).$$

The theorem follows from this.

The above relation between integral points and Hilbert's theorem had already been pointed out by Siegel in his first paper [Si1].

S-integral points.

We take the ground field to be $\mathbf{Q}$. Take $\alpha \in \mathbf{Q}$ and $S = \{p_1, ..., p_l\}$ a finite set of prime numbers.

We write

$$Q_{\alpha,S} = \{t \in \mathbf{Q}; t - \alpha \text{ is of the form } \pm \prod_{i=1}^{l} p_i^{m_i}, m_i \in \mathbf{Z}\}.$$

In particular, for $\alpha = 0$, $Q_{0,S}$ is the set of units of

$$\mathbf{Z}_S = \mathbf{Z}[\frac{1}{p_1}, ..., \frac{1}{p_l}] = \mathbf{Z}[\frac{1}{p_1 ... p_l}].$$

If $S \ne \emptyset$, $Q_{\alpha,S}$ is an infinite set. It is a quasi-integral set for $\mathbf{P}_1 - \{\alpha, \infty\}$ over the ring $\mathbf{Z}_S$.

Theorem. *Let Ω be a thin set in* $\mathbf{Q}$ *and* S *a finite set of primes. For* α *outside a finite set,* $\Omega \cap Q_{\alpha,S}$ *is finite.*

Proof.

We may assume that Ω is contained in $\pi(X(\mathbf{Q}))$, where π, X are as above. We choose α such that $\pi^{-1}(\alpha)$ has at least two elements (for example, we take α different from the ramification points, since the degree of π is ≥ 2). We take $X_\infty = \pi^{-1}(\infty \cup \{\alpha\})$. Then X_∞ has at least 3 points. From the Siegel-Mahler-Lang theorem, every quasi-integral set with respect to $\mathbf{Z}_S$ over $X_{\text{aff}} = X - X_\infty$ is finite, hence the theorem.

Exercises. 1). Generalise the above theorem to arbitrary number fields.
2). Show that every thin subset of $\mathbf{Z}$ is contained in the finite union of sets of the following type:

a) a finite set,
b) the intersection of $\mathbf{Z}$ with the set of values of $\phi(x, y)$, where ϕ is a rational function with coefficients in $\mathbf{Q}$ and x, y run through the solutions of some Pell equation $x^2 - Dy^2 = 1$ (where $D > 0$ is a non-square).
c) the intersection of $\mathbf{Z}$ with the image of a polynomial of degree $\delta \geq 2$ with rational coefficients.

If Ω is of type (b), resp. (c), show that $N_\Omega(B)$ is of order of magnitude $\log B$, resp. $B^{1/\delta}$.

10. CONSTRUCTION OF GALOIS EXTENSIONS

The following conjecture is part of the folklore.

If G is a finite group, there is a Galois extension F of $\mathbf{Q}$ whose Galois group is G.

This applied to $G \times G \times \ldots \times G$ would imply that there are infinitely many disjoint extensions F of $\mathbf{Q}$ with $\mathrm{Gal}(F/\mathbf{Q}) = G$; the same statement for number fields would then follow.

This conjecture is known when G is solvable, as proved by Shafarevich, cf. [Sh], [Is] (both proofs [Sh] and [Is] contain a mistake relative to the prime 2, but this is believed to be repairable); for more complete results in the case of odd order, see [Ne].

As we will see, Hilbert's irreducibility theorem can be used to handle other cases.

10.1. The method.

Theorem. *Let K be a hilbertian field and G a finite group. If there exists a Galois extension $F/K(T_1, \ldots, T_n)$, with Galois group G, there is such an extension for K. Moreover, if F is K-regular (i.e. K is algebraically closed in F), there exist infinitely many such extensions which are linearly disjoint.*

Proof. The first assertion follows directly from proposition 2 of §9.2. Recall that the extension K is obtained by "specialising" $\underline{T} = (T_1, \ldots, T_n)$ to $\underline{t} = (t_1, \ldots, t_n) \in K^n$, and that the set of $\underline{t}$'s which give a Galois extension $F_{\underline{t}}/K$ with Galois group G is the complement of a thin set Ω_K.

Assume now that F is K-regular, and that we have constructed m linearly disjoint Galois extensions $F_1, \ldots, F_m$ of K with Galois group G. Let K' be the composite of $F_1, \ldots, F_m$. Since F is K-regular, the extension $K'F/K'(T_1, \ldots, T_n)$ still has Galois group G. Applying the proposition in §9.4, we see that $K^n \cap \Omega_{K'}$ is thin in K^n. Since K is hilbertian, we can choose $\underline{t} \in K^n$, with $\underline{t} \notin \Omega_{K'}$ and the corresponding extension $F_{\underline{t}}/K$ has Galois group G, and is linearly disjoint from K'. By adding it to $F_1, \ldots, F_m$ we get $m+1$ extensions of the required type. The theorem follows by induction on m.

Corollary. *Assume that G has the property:*

(Gal_T) - *There exists a regular Galois extension of $\mathbf{Q}(T)$ with Galois group G.*

Then every hilbertian field of characteristic zero has an infinite number of Galois extensions with Galois group G.

Proof.
If $E/\mathbf{Q}(T)$ is regular with Galois group G, and K is any field of characteristic zero, then $F = T \otimes_{\mathbf{Q}(T)} K(T)$ is a regular Galois extension of $K(T)$ with Galois group G. One then applies the theorem to F.

Remark. An optimistic conjecture is that Gal_T is true for every finite group G; this is stronger than the conjecture above.

Exercises. 1) Prove that Gal_T is true for all finite abelian groups (whether this extends to solvable groups is unknown).
2) If Gal_T is true for two groups G_1 and G_2 then it is true for $G_1 \times G_2$.
[Use an automorphism of $\mathbf{Q}(T)$ to make the ramification sets disjoint.]

10.2. Extensions with Galois group S_n.

A first example is to apply the method to the generic equation of degree n:

$$X^n + T_1 X^{n-1} + \ldots + T_n = 0,$$

which has Galois group S_n over $K(T_1, \ldots, T_n)$. Hence if K is hilbertian, it has infinitely many disjoint extensions with Galois group S_n.

A more interesting example is that given in Hilbert's original paper [Hi], namely:

Theorem. *Let $f(X) = X^n + a_1 X^{n-1} + \ldots + a_n$, $n \geq 2$, be a polynomial over a field K of characteristic 0, such that*
a) the roots $\beta_1, \ldots, \beta_{n-1}$ of f' are simple,
b) the values of f at the points β_i are distinct.
Then the equation $f(X) - T = 0$ is irreducible over $K(T)$ and has Galois group S_n.

Corollary. *There is a thin set $\Omega \subset K$ such that for every $t \in K - \Omega$, the equation $f(X) - t = 0$ has Galois group S_n. Further, if K is hilbertian* *then there are infinitely many linearly disjoint extensions of K with Galois group S_n.*

For example over $\mathbf{Q}$, $X^n - X - t = 0$ has Galois group S_n for all t outside a thin set, in particular for infinitely many prime numbers t.

We shall need a lemma on finite groups.

Lemma 1. *Let n be an integer ≥ 1, and let G be a subgroup of S_n. Assume that*

i) G is transitive
ii) G contains a transposition
iii) G is generated by cycles of prime order.
Then $G = S_n$.

Proof.
Let E be a finite set of n elements, S_E the symmetric group of E, and G a subgroup of S_E with the properties (i),(ii) and (iii). Let us show by induction on m that if m is an integer $\leq n$, there is a subset E_m of E with m elements such that $G \supset S_{E_m}$ (where S_{E_m} is identified with a subgroup of S_E with trivial action on $E - E_m$). For $m = n$ this will show that $G = S_n$.
For $m = 1$ there is nothing to prove. To make the step from m to $m + 1$, where $1 \leq m < n$, we consider two cases.

(a) *There is a transposition $\tau = (xy) \in G$ with $x \in E_m$, $y \notin E_m$.*

In this case we select $E_{m+1} = E_m \cup \{y\}$. Let H be the group generated by S_{E_m} and τ. It is clear that H is a subgroup of G which stabilizes E_{m+1}, that H fixes $E - E_{m+1}$, that H acts transitively on E_{m+1}, and that H contains S_{E_m}. The index of S_{E_m} in H is at least $m + 1$ therefore $H = S_{E_{m+1}}$, and we have shown that G contains $S_{E_{m+1}}$.

(b) *If a transposition $\tau = (xy)$ belongs to G, then either x and $y \in E_m$ or x and $y \in E - E_m$.*

Let N be the subgroup generated by the transpositions of G. By (b), N stabilises E_m, and there are at least two orbits of N on E. We decompose E into orbits under the action of N:

$$E = F_1 \cup ... \cup F_k,$$

with $k \geq 2$, and the F_i are pairwise disjoint. The subgroup N is normal in G, consequently the F_i are permuted by G; but G is transitive, therefore the F_i are permuted transitively by G, and have the same number of elements. From hypothesis (ii), the F_i are not all reduced to 1 point, therefore

$$|F_1| = ... = |F_k| \geq 2.$$

Let $\sigma = (x_1 \ldots x_p)$ be a cycle of prime order p belonging to G. Let us show that $\sigma F_i = F_i$ for all i. Indeed, if i_1 is an index such that $\sigma F_{i_1} \neq F_{i_1}$, define $i_2, \ldots, i_p$ by $\sigma F_{i_k} = F_{i_{k+1}}$, $(1 \leq k < p)$ and since p is prime, the indices $i_1, i_2, \ldots, i_p$ are distinct. Let x and y be two distinct elements of F_{i_1}; the orbits under $\{1, \sigma, \ldots, \sigma^{p-1}\}$ of x and y are distinct, and of order p, which is impossible for a cycle.

From hypothesis (iii), we have $\sigma F_i = F_i$ for all $\sigma \in G$ and all i, which contradicts the fact that the F_i are permuted transitively by G with $k \geq 2$. Hence case (b) is impossible and this concludes the proof of the induction step.

We go back to Hilbert's example: let f be a polynomial of degree n over K, such that the roots $\beta_1, \ldots, \beta_{n-1}$ of f' are simple, and the $f(\beta_i)$ are pairwise distinct. We wish to show that the equation $f(X) - T = 0$ is irreducible with Galois group S_n for generic T, therefore also for $T \in K - \Omega$ with Ω thin.

We map the projective line to itself by f which is a morphism of degree n, and $K(X,T)$ is an extension of $K(T)$ of degree n (the irreducibility of the polynomial $f(X) - T$ over $K(T)$ is clear). Let F be the Galois closure of $K(X,T)$ over $K(T)$ and let $G \subset S_n$ be the Galois group of F over $K(T)$, and $H = G \cap S_{n-1}$ the Galois group of F over $K(X,T)$.

$$G \left(\begin{array}{c} F \\ \uparrow \\ K(X) \\ \uparrow f \\ K(T) \end{array} \right. \quad \left. \begin{array}{c} \\ \\ \\ \end{array} \right) H$$

The homogeneous space G/H may be identified with the set of roots of the equation $f(X) - T = 0$.

We use a lemma on coverings of the projective line. Over a field of characteristic 0 (that we may assume is algebraically closed; we may even assume that it is $\mathbf{C}$), we consider a Galois covering $\pi : X \to \mathbf{P}_1$, with X irreducible. Let G be the Galois group.

Lemma 2. *Let $Q \in \mathbf{P}_1$. The inertia groups of the points $x \in X$ for which $\pi(x) \neq Q$ generate the group G.*

Taking the subgroup generated by these inertia groups, and passing to the quotient, this reduces to the following lemma.

Lemma 3. *If a Galois covering $X \to \mathbf{P}_1$ is unramified except at one point Q, then it has degree 1.*

This is clear over $\mathbf{C}$, because $\pi_1(\mathbf{P}_1 - \{Q\}) = \{1\}$ (the affine line is simply connected).

[More generally, Riemann's existence theorem allows one to describe the Galois coverings of $\mathbf{P}_1$ (over $\mathbf{C}$, or over any algebraically closed field of characteristic 0) which are unramified outside a given set of points $\{Q_1, ..., Q_m\}$: the Galois group G of such a covering can be generated by m elements $\sigma_1, ..., \sigma_m$ (which are generators of some inertia subgroups above the Q_i's), with the relation $\sigma_1 ... \sigma_m = 1$.

Conversely, any finite group G with such generators can be obtained in this way. This follows from the structure of the fundamental group of a sphere with m punctures.]

We can now complete the proof of the theorem.

The zeros of f' are $\beta_1, ..., \beta_{n-1}$. Hence the covering $X \to \mathbf{P}_1$, corresponding to the field extension $F/K(T)$, is unramified outside $Q = \infty$ and $f(\beta_1), ..., f(\beta_{n-1})$. Using the fact that the $f(\beta_i)$ are distinct, one sees that the inertia subgroups G_P of G at a point P above one of the β_i is cyclic of order 2 and generated by a transposition. Since G is generated by such subgroups, lemma 1 shows that G is equal to S_n.

Complements. Here is some more information on this example. Assume K is a number field, and let S be a finite set of places of K containing the archimedean places, and A the ring of S-integers of K. Let $\Omega_{f,A}$ be the set of $t \in A$ such that $f(X) - t$ does not have S_n for its Galois group. It is clear that if t is a value of f, then $t \in \Omega_{f,A}$:

$$\Omega_{f,A} \supset A \cap f(K).$$

Theorem. *If $n \geq 5$, then $\Omega_{f,A}$ is the union of $A \cap f(K)$ and a finite set.*

The proof of finiteness will unfortunately be ineffective.

Corollary. *Given $n \geq 5$, the equation $X^n - X - p = 0$ has Galois group S_n over $\mathbf{Q}$ for all prime numbers p except a finite number.*

Indeed, no prime number $p > 2$ is of the form $x^n - x$, $x \in \mathbf{Q}$, therefore for $f(X) = X^n - X$ and $A = \mathbf{Z}$, $\Omega_{f,A}$ contains only a finite number of prime numbers.

Proof of theorem.

Consider the curve X attached to f as before; it is a Galois covering of $\mathbf{P}_1$, with Galois group S_n.

$$\left.\begin{array}{c} X \\ \downarrow \\ \mathbf{P}_1 \end{array}\right) G = S_n$$

If t belongs to $\Omega_{f,A}$. then either t is equal to one of the $f(\beta_i)$ (in which case $f(X) - T$ has a double root) or t can be lifted to a rational point of some X/H where H is a subgroup of S_n distinct from S_n. The subgroup S_{n-1} (and its conjugates) corresponds to $A \cap f(K)$. We have to show that the others only give a finite set. This follows (via Siegel's theorem) from the following lemma.

Lemma 4. *If $n \geq 5$, and if H is a subgroup of S_n distinct from S_n, and not conjugate to S_{n-1}, then the covering $X/H \to \mathbf{P}_1$ is not of exceptional type, i.e. it is not of genus 0 with at most two points at infinity.*

Proof.
a) If H is the alternating group A_n, $X/H \to \mathbf{P}_1$ is quadratic, and ramified at $n-1$ points or at n points according to whether n is odd or even. As $n \geq 5$, this gives an hyperelliptic curve of genus ≥ 1.
b) In the other cases, the representation of S_n in S_m (where $m = [G : H]$) attached to H is faithful (use $n \geq 5$ and the simplicity of A_n). This implies in particular $m \geq n$ (for $n!$ must divide $m!$) and even $m > n$ if $n \neq 6$ (for the automorphisms of S_n are inner for $n \neq 6$).

Let $\rho : S_n \to S_m$ be this representation. The image by ρ of a transposition is an element of order 2, i.e. a product of p transpositions ($p \geq 1$), fixing $m - 2p$ elements. On the other hand, a cycle of degree n gives by ρ a product of cycles of degrees $n/d_1, ..., n/d_h$, with $m = \sum n/d_i$, $\gcd(d_i) = 1$. As the inertia at infinity is cyclic of order n, if X/H is of exceptional type then $h \leq 2$ since h is the number of points at infinity in X/H.

On the other hand, Hurwitz' formula for the genus gives

$$2g - 2 = -2m + (n-1)p + \sum_{i=1}^{h} \left(\frac{n}{d_i} - 1\right),$$

where g is the genus of X/H. In the exceptional case we have $g = 0$, $h = 1$ or 2, whence

$$m = (n-1)p + 2 - h.$$

i) If $h = 1$, we have $m = n/d_1$, and as $m \geq n$ this implies $d_1 = 1$, $m = n$, whence $m = n = 6$, and ρ is a non-inner automorphism $S_6 \to S_6$. This automorphism transforms a transposition (ab) to a product of three transpositions: $(ab)(cd)(ef)$. We have thus $p = 3$, and the equation $m = (n-1)p + 1$ is violated.
ii) If $h = 2$, we have $m = n/d_1 + n/d_2$, whence $m \leq 2n$. As $m = (n-1)p$ this gives $p \leq 2n/(n-1) < 3$, whence $p = 1$ or 2. If $p = 1$ we have $m = n - 1 < n$, impossible. Therefore $p = 2$ and $m = 2(n-1)$. But then d_1 and d_2 are not both equal to 1: one of them is ≥ 2, and $n/d_1 + n/d_2 \leq \frac{3}{2}n$. Whence $2(n-1) = m \leq \frac{3}{2}n$ and $n \leq 4$.

Remark. For $n = 4$ and $H = D_4$, the dihedral group of order 8, one finds that X/H has genus 0 and there are two points at infinity. This corresponds with the above notation to $m = 3$, $p = 1$, $h = 2$, $d_1 = 2$, $d_2 = 4$. In this case $S_n \to S_m$ is not faithful.

Exercises. 1) Give an algebraic proof of lemma 3, by using Hurwitz' formula for the genus of a ramified covering. Show that the results holds in characteristic $p > 0$ provided the covering is tamely ramified at Q.
2) Let E be a finite set with $n \geq 1$ elements and G a subgroup of S_E. Define a relation R on $E \times E$ by

$$R(x, y) = \{x = y \text{ or the transposition } (xy) \text{ belongs to } G\}.$$

a) Prove that R is an equivalence relation on E (use $(xz) = (xy)(yz)(xy)$ if x, y, z are distinct elements of E).
b) Let F be an equivalence class mod R. Show that S_F is contained in G.
c) Let $F_1, ..., F_m$ be different equivalence classes modulo R. Prove that $S_{F_1} \times ... \times S_{F_m}$ is a normal subgroup of G.
d) Let $\tilde{G}$ be the group of permutations of $\tilde{E} = E/R$ defined by G. Show that $s \in S_E$ belongs to G if and only if it is compatible with R and its image in $S_{\tilde{E}}$ belongs to $\tilde{G}$.
e) Use d) to recover lemma 1.

3) Let $F(X) = X^n + a_1X^{n-1} + ... + a_n$ be a polynomial with coefficients in $\mathbf{Z}$, with distinct roots (in $\bar{\mathbf{Q}}$). Assume that, for every prime number p, the reduction $\tilde{f}$ of f mod p has either n distinct roots (i.e. p does not divide discr(f)), or it has $n-2$ distinct simple roots and one double root. Prove that the Galois group G of f (viewed as a subgroup of S_n) is generated by transpositions (use the fact that G is generated by the inertia subgroups, and show that these are either trivial or generated by a transposition).

Conclude that, if f is irreducible, then $G = S_n$ and the A_n extension of $\mathbf{Q}(\sqrt{\mathrm{discr}(f)})$ thus obtained is unramified.
4) Let $n \geq 2$. Using the fact (proved by E. Selmer, Math. Scand. (1956), 287-302) that $X^n - X - 1$ is irreducible over $\mathbf{Q}$, show that its Galois group is S_n (use exercise 3). Give a direct proof for $n = 2, 3, 4, 5$.

10.3. Extensions with Galois group A_n.

Here is another example of Hilbert.

Let n be an even integer ≥ 4. Consider a polynomial $g(X)$ of degree n, divisible by X^2, such that

$$g'(X) = nX(X - \beta_1)^2 ... (X - \beta_h)^2,$$

where $h = n/2 - 1$. Assume that the $\beta_i \in K$ are distinct and non-zero, and the $g(\beta_i)$ are distinct and non-zero.

Theorem. *For generic t (thus also for t in K outside a thin set) the Galois group of the equation $g(X) + (-1)^{n/2}t^2 = 0$ is A_n.*

For n odd, Hilbert's paper gives a similar equation also leading to the alternating group A_n.

Proof.

We begin by showing that the Galois group of $F(X) = g(X) - T$ is S_n. The proof is the same as before; the inertia groups over the β_i are generated by cycles of order 3, and at 0 by a transposition. Then G is generated by these elements, and as G is transitive we obtain $G = S_n$. Hence the Galois extension Ω_F generated over $k(T)$ by the roots of $F(X)$ has Galois group S_n.

We now factorise $F(X) = g(X) - T$ in the algebraic closure:

$$F(X) = \prod_{i=1}^{n} (X - \alpha_i),$$

where the discriminant is:

$$\mathrm{discr}\ F = \prod_{i<j} (\alpha_i - \alpha_j)^2.$$

Eliminating X between F and its derivative, one finds the classical formula

$$\mathrm{discr}\ F = n^n (-1)^{n(n-1)/2} \prod_{j=1}^{n-1} F(\gamma_j),$$

where $\gamma_1, ..., \gamma_{n-1}$ are the roots of F'. Here

$$\{\gamma_1, ..., \gamma_{n-1}\} = \{0, \beta_1, \beta_1, ..., \beta_h, \beta_h\},$$

and the discriminant of F is the product of $(-1)^{1+n/2}T$ with a square.

Hence, the subfield of Ω_F corresponding to A_n is the field $K(T, \sqrt{\mathrm{discr} F}) = K(t)$, where t is a square root of $(-1)^{1+n/2}T$. This shows that the Galois group of the equation $g(X) + (-1)^{n/2}t^2$ is indeed A_n.

Schur has given some numerical examples of equations with Galois group S_n or A_n ("Affektlöse Gleichungen"). For example, the Galois group of $1 + X + ... + X^n/n! = 0$ over $\mathbf{Q}$ is S_n if $4 \nmid n$ and A_n if $4|n$ (see [Schu]).

10.4. Further examples of Galois groups: use of elliptic curves.

Let E be an elliptic curve over a field K, and n an integer ≥ 1. The kernel of multiplication by n gives a Galois extension whose Galois group $G_{E,n}$ embeds in the group $\mathrm{GL}(2, \mathbf{Z}/n\mathbf{Z})$.

Over $\mathbf{Q}$, Hilbert's theorem shows that $G_{E,n}$ is "usually" equal to $\mathrm{GL}(2, \mathbf{Z}/n\mathbf{Z})$, but does not provide any specific example.

Over the field $\mathbf{Q}(t)$, where t is an indeterminate, we take an elliptic curve with modular invariant j equal to t e.g.

$$y^2 + xy = x^3 - \frac{36}{t - 1728}x - \frac{1}{t - 1728}.$$

For all n, the Galois group $G_{E,n}$ over $\mathbf{Q}(t)$ is the group $\mathrm{GL}(2, \mathbf{Z}/n\mathbf{Z})$.

Indeed, we remark first that the image $\tilde{G}_{E,n}$ in $\mathrm{GL}(2, \mathbf{Z}/n\mathbf{Z})/(\pm 1)$ is canonical (independent of the choice of E with $j = t$). The function field of the curve which classifies the n-division points is the field of modular functions of level n with coefficients in the cyclotomic field $\mathbf{Q}(\sqrt[n]{1})$. The theory of modular functions implies $\tilde{G}_{E,n} \doteq \mathrm{GL}(2, \mathbf{Z}/n\mathbf{Z})/(\pm 1)$. Finally, if H is a subgroup of $\mathrm{GL}(2, \mathbf{Z}/n\mathbf{Z})$ such that $H.(\pm 1) = GL(2, \mathbf{Z}/n\mathbf{Z})$, then $H = GL(2, \mathbf{Z}/n\mathbf{Z})$. (If not, H would have index 2, and as $\sigma = \begin{pmatrix} 0 & -1 \\ 1 & 0 \end{pmatrix}$ satisfies $\sigma^2 = -1$ we would have $\sigma^2 = -1 \in H$, whence a contradiction.)

Let K_n be the field generated by the n-division points of the generic curve, $K_{n,t}$ that of the corresponding curve E_t:

$$GL_2(\mathbf{Z}/n\mathbf{Z})\left(\begin{array}{c} K_n \\ \uparrow \\ \mathbf{Q}(\sqrt[n]{1})(t) \\ \uparrow \\ \mathbf{Q}(t) \end{array}\right. \begin{array}{l} \left.\begin{array}{c} \\ \\ \end{array}\right\} SL_2(\mathbf{Z}/n\mathbf{Z}) \\ \left.\begin{array}{c} \\ \\ \end{array}\right\} GL_1(\mathbf{Z}/n\mathbf{Z}) = (\mathbf{Z}/n\mathbf{Z})^* \end{array}$$

We have

$$\mathrm{Gal}(K_n/\mathbf{Q}(t)) = GL_2(\mathbf{Z}/n\mathbf{Z}),$$

$$\mathrm{Gal}(K_n/\mathbf{Q}(\sqrt[n]{1})(t)) = SL_2(\mathbf{Z}/n\mathbf{Z}),$$

$$\mathrm{Gal}(\mathbf{Q}(\sqrt[n]{1})(t)/\mathbf{Q}(t)) = (\mathbf{Z}/n\mathbf{Z})^*.$$

For all n, there is a thin set $\Omega_n \subset \mathbf{Q}$ (depending on n) such that for $t \notin \Omega_n$, the corresponding curve E_t satisfies $\mathrm{Gal}(K_{n,t}/\mathbf{Q}) = \mathrm{GL}_2(\mathbf{Z}/n\mathbf{Z})$. (When $\mathbf{Q}$ is replaced by a number field K one obtains a similar assertion, assuming that the extensions $K/\mathbf{Q}$ and $\mathbf{Q}(\sqrt[n]{1})/\mathbf{Q}$ are disjoint.)

Example. For the curve $y^2 - y = x^3 - x$ of conductor 37, one can prove that $G_{E,n} = \mathrm{GL}(2, \mathbf{Z}/n\mathbf{Z})$ for all n not divisible by $74 = 2 \times 37$. Nevertheless, there is no elliptic curve for which $G_{E,n} = \mathrm{GL}(2, \mathbf{Z}/n\mathbf{Z})$ for all n (see [Se6, §5.5, Prop. 22]).

Let us now search for Galois extensions of $\mathbf{Q}$ with Galois group the simple group $\mathrm{PSL}_2(\mathbf{F}_p)$, where p is a prime ≥ 5.

If, instead of $\mathbf{Q}$, we were interested in $\mathbf{Q}(\sqrt{(-1)^{\frac{p-1}{2}}p})$, we could use the $\mathrm{GL}_2(\mathbf{Z}/p\mathbf{Z})$-extensions of $\mathbf{Q}$ constructed above. This does not work over $\mathbf{Q}$. Nevertheless, the following result has been proved by K-y. Shih ([Sh1], [Sh2]):

Theorem. *Let p be an odd prime number such that $(\frac{2}{p}) = -1$, or $(\frac{3}{p}) = -1$, or $(\frac{7}{p}) = -1$. Then there is a regular Galois extension of $\mathbf{Q}(T)$ with Galois group $PSL_2(\mathbf{F}_p)$.*

The proof is roughly as follows. Let $p^* = (-1)^{\frac{p-1}{2}}p$ and $L_p = \mathbf{Q}(\sqrt{p^*})$. Let M_N (where N is any positive integer) be the function field $\mathbf{Q}(j, j_N)$ of the modular curve $X_0(N)$ (recall that $j_N(z) = j(Nz) = j(-1/Nz)$); let w

be the canonical involution of M_N (which permutes j and j_N). Call $M_{N,p}$ the "twisted form" of M_N relative to $L_p/\mathbf{Q}$, i.e. the subfield of $L_p \otimes M_{N,p}$ fixed by $\sigma \otimes w$, where σ is the involution of L_p. (This field has a modular interpretation in terms of N-isogenies of a curve E onto its σ-conjugate E^σ.) Using this modular interpretation, Shih shows:
a) if $(\frac{N}{p}) = -1$, there is a regular Galois extension of $M_{N,p}$ with Galois group $\mathrm{PSL}_2(\mathbf{F}_p)$;
b) if N =2,3, or 7, $M_{N,p}$ is a pure transcendental extension of $\mathbf{Q}$ (i.e. it corresponds to a curve of genus 0 which has a rational point).

For more details see Shih *loc. cit.*

This applies in particular to $p = 7$, where $\mathrm{PSL}_2(\mathbf{F}_p)$ has order 168 and is isomorphic to $\mathrm{SL}_3(\mathbf{F}_2)$. In that case, another construction of a regular extension of $\mathbf{Q}(T)$ with that Galois group has been given by LaMacchia ([LaM], see also [MM]); an especially interesting extension of $\mathbf{Q}$ obtained in this way is the field of roots of $x^7 - 7x + 3 = 0$.

10.5. Noether's method.

Here is a tentative method suggested by Emmy Noether for constructing extensions with given Galois group G. We make G operate faithfully on a finite set E. We consider the field $\mathbf{Q}(T_x, x \in E)$ where T_x are indeterminates indexed by E, and make G operate by $T_x \mapsto T_{\sigma x}$ for $\sigma \in G$. Let K_G be the field of invariants. If K_G is a purely transcendental extension of $\mathbf{Q}$, the group G can be realised as a Galois group by §10.1.

The simplest try consists of choosing $E = G$, and making G operate on itself by translation. But Swan has shown that even for cyclic extensions, K_G may not be pure. The first example of Swan was $G = \mathbf{Z}/47\mathbf{Z}$. Lenstra has analysed this question and his simpler example is $\mathbf{Z}/8\mathbf{Z}$ [Len].

[It thus appears that Noether's method does not work, at least in the crude form described above. Nonetheless, one can hope that the field K_G, without being pure, still has enough good properties for some form of the irreducibility theorem to apply, cf. Ekedahl [E].]

If we take $G = A_n$ operating on n letters, K_G is the field generated by the elementary symmetric functions $\sigma_1, ..., \sigma_n$ and by the square root of the discriminant. For $n = 2, 3, 4$, K_G is a purely transcendental extension. One would like to know what happens for other values of n.

10.6. Infinite Galois extensions.

We now extend Hilbert's irreducibility theorem to some infinite Galois extensions. We need first an auxiliary result on profinite groups.

Proposition. *Let G be a profinite group. The following properties are equivalent:*

(i) There is an open normal subgroup U of G such that $H.U \neq G$ for every closed normal subgroup H of G with $H \neq G$.

(ii) The Frattini subgroup $\Phi(G)$ of G is open.

(iii) The prime numbers dividing the order of G are finite in number, and the Sylow subgroups of G are (topologically) finitely generated.

(iv) There is an open normal subgroup N of G with $N = \prod_{p \in P} N_p$, where P is a finite set of primes, and each N_p is a finitely generated pro-p-group.

(Recall that the Frattini subgroup $\Phi(G)$ of G is the intersection of the maximal closed proper subgroups of G, just as in the finite case, cf. [Hup], Kap. III,§3. One has $\Phi(G) = \varprojlim \Phi(G/V)$, where V runs through the open normal subgroups of G.)

Proof.

The equivalence (i) $\Leftrightarrow$ (ii) is clear; so is the implication (iv) $\Rightarrow$ (iii). Note that if P is a pro-p-group, $\Phi(P) = (P.P)^{\mathrm{cl}}.P^p$ is the intersection of the kernels of the continuous homomorphisms $P \to \mathbf{Z}/p\mathbf{Z}$; hence $\Phi(P)$ is open if and only if P is (topologically) finitely generated.

Let us now assume (iii) and prove (i). Let $P_1, ..., P_h$ be Sylow subgroups of G corresponding to the primes $p_1, ..., p_h$ dividing the order of G. By assumption, $\Phi(P_i)$ is open in P_i. Hence there is a normal open subgroup U of G such that $U \cap P_i \subset \Phi(P_i)$ for all i. We show that U has property (i). Note first that the Sylow subgroups are conjugate and that U is invariant under conjugation, thus the property $U \cap P_i \subset \Phi(P_i)$ is satisfied for all the Sylow p_i-subgroups.

Let H be a closed subgroup such that $H.U = G$. Let $H_1, ..., H_h$ be Sylow subgroups of H for the different primes $p_1, ..., p_h$. We can select the P_i so that $H_i \subset P_i$. As H maps onto G/U, H_i maps onto a Sylow p_i-subgroup of G/U, thus the images of H_i and P_i in G/U coincide:

$$P_i = H_i.(U \cap P_i).$$

As $U \cap P_i \subset \Phi(P_i)$ (by construction), we find

$$P_i = H_i.\Phi(P_i)$$

therefore $P_i = H_i$. But a group is generated by a choice of its Sylow p-subgroups for each p (this is true by Sylow's theorem for finite groups, and this extends to profinite groups). Whence $H = G$.

It remains to prove that (ii) $\Rightarrow$ (iv). Since $N = \Phi(G)$ is pro-nilpotent ([Hup], p.270, Satz 3.6), it can be written as a direct product $N = \prod_{p \in P} N_p$, where the N_p are pro-p-groups. By assumption N is open in G; if S is a finite set of elements generating G and N, then S generates G (topologically); this shows that G, and hence N and the N_p, are finitely generated. If p is a prime which does not divide $[G : N]$, then p does not divide the order of N (see [Hup], p.270, Satz 3.8); this shows that $N_p = 1$ for almost all p, and concludes the proof.

Examples. 1) A finite product of compact p-adic Lie groups has properties (i)-(iv) above (property (iv) is especially easy to check).
2) The group $\hat{\mathbf{Z}} = \prod_p \mathbf{Z}_p$ does *not* have these properties.

Profinite groups G having properties (i)-(iv) behave as well as finite groups for Hilbert's irreducibility theorem.

Namely, let G be such a group, and $F/K(T)$ a Galois extension with Galois group G. Assume:

(Ram) There is a finite set S of places of $K(T)$ outside of which F is unramified.

Then if $t \in K$ does not belong to S, the decomposition subgroup G_t of t is the Galois group of the corresonding residue field extension F_t/K; it is a subgroup of G (defined up to conjugation).

Theorem. *There is a thin set $\Omega \subset K$ such that $G_t = G$ for all $t \in K - \Omega$.*

Proof.
Let $N = \Phi(G)$ be the Frattini subgroup of G and let $F^N/K(T)$ be the corresponding extension of $K(T)$. The Galois group of $F^N/K(T)$ is G/N, which is finite by assumption. By Proposition 2 of §9.2, there is a thin set Ω of K such that, if $t \notin \Omega \cup S$, the decomposition group $(G/N)_t$ is equal to G/N. For such a t we have $G_t.N = G$, hence $G_t = G$ since N is the Frattini subgroup of G.

Exercise. Use the theorem above to construct extensions of $\mathbf{Q}$ with Galois group $\mathrm{GL}_2(\mathbf{Z}_l)$ for any given l.

[10.7. Recent results.

There has been much progress in the 1980's towards realising a given group as a Galois group over $\mathbf{Q}$, or over $\mathbf{Q}(T)$. For a detailed exposition (resp. a résumé), see Matzat's book [Mat] (resp. the Bourbaki report [Se5]).

One of the main results concerns groups having a *rigid family of rational conjugacy classes.* Let us define "rational" and "rigid":

A conjugacy class C of a finite group G is called *rational* if the following equivalent properties hold:

(a) every character of G takes rational values on C;

(b) if $g \in C$, one has $g^m \in C$ for every integer m which is prime to the order of C.

Let now $C_1, \dots, C_k$ be conjugacy classes of G. Assume G has trivial centre. Define $P = P(C_1, \dots, C_k)$ as the set of all $(g_1, \dots, g_k) \in C_1 \times \dots \times C_k$ such that $g_1 \dots g_k = 1$; define P' as the subset of P made of those $(g_1, \dots, g_k) \in P$ which generate G. The group G acts by conjugation on P and P': an element $g \in G$ transforms $(g_1, \dots, g_k)$ into $(gg_1g^{-1}, \dots, gg_kg^{-1})$. One says that $(C_1, \dots, C_k)$ is *rigid* if $P' \neq \emptyset$ and G acts transitively on P' (i.e. P'/G has one element).

Note that G acts freely on P'. Hence the rigidity property is equivalent to $|P'| = |G|$, and it implies $|P| \geq |G|$. (This is useful, because $|P|$ can be computed from the character table of G:

$$|P| = \frac{1}{|G|} \prod_{i=1}^{k} |C_i| \sum_{\chi} \chi(c_1) \dots \chi(c_k) / \chi(1)^{k-2},$$

where c_i is a representative element of C_i and χ runs through the irreducible characters of G).

Theorem. (Belyi-Fried-Matzat-Thompson). *Let G be a finite group with trivial centre, and $C_1, \dots, C_k$ be conjugacy classes of G. Assume that each C_i is rational and that the family $(C_1, \dots, C_k)$ is rigid. Then there exists a regular Galois extension of $\mathbf{Q}(T)$ with Galois group G.*

Hence G has property Gal_T of §10.1. In particular, G can be realised as a Galois group over every hilbertian field of characteristic zero.

The *proof* gives a more precise result. One selects k distinct rational points $t_1, \dots, t_k$ of the projective line $\mathbf{P}_1$, and one constructs a regular Galois covering $X \to \mathbf{P}_1$ over $\mathbf{Q}$ with Galois group G having the following properties:

(1) it is unramified outside $t_1, \dots, t_k$;

(2) for each i, $1 \leq i \leq k$, the inertia group in G relative to t_i (which is well defined up to conjugation) is generated by an element of C_i.

(Idea of the construction : one does it first over $\mathbf{C}$, using the explicit description of the fundamental group of a sphere with k punctures; then, one shows that the covering so obtained is unique, up to unique isomorphism; by descent theory this implies that its field of definition can be lowered from **C** to **Q**.)

The condition that G contains a rigid rational family of conjugacy classes is rather restrictive. It can be weakened in several ways (see [Mat], [Se5]). Using the variants so obtained, the property Gal_T has been proved[(1)] for the following groups:

A_n, S_n;
$\mathrm{PSL}_3(\mathbf{F}_p), p \equiv 1 \pmod 4$;
$\mathrm{PSp}_4(\mathbf{F}_p), p \geq 3, p \equiv 2, 3 \pmod 5$;
$\mathrm{G}_2(\mathbf{F}_p), p \geq 5$;
the sporadic groups M_{11}, M_{12}, J_1, M_{22}, J_2, HS, M_{24}, He, Suz, ON, Co_3, Co_2, Fi_{22}, HN, F_3, Fi_{23}, Co_1, J_4, Fi'_{24}, $\mathrm{F}_2 = \mathrm{BM}$ and $\mathrm{F}_1 = \mathrm{M}$.

Explicit equations have been given for some groups: $\mathrm{PSL}_2(\mathbf{F}_7)$, $\mathrm{PSL}_2(\mathbf{F}_{11})$, $\mathrm{PSL}_2(\mathbf{F}_{13})$, $\mathrm{SL}_2(\mathbf{F}_8)$, M_{12}.

As for groups with non-trivial centre, one may mention $\tilde{A}_n = 2.A_n$ (non-trivial extension of A_n by a group of order 2): it has been proved by N. Vila (see e.g. [Mat], Kap. IV, §6) that $\tilde{A}_n$ has property Gal_T when $n \equiv 0, 1 \pmod 8$. For other classes mod 8, only partial results are known.]

---

(1) In some cases, the proof assumes the classification of finite simple groups.

11. CONSTRUCTION OF ELLIPTIC CURVES OF LARGE RANK

11.1. Néron's specialisation theorem.

Let k be a number field and A an abelian variety over $K = k(T_1, ..., T_n)$ where the T_i are indeterminates over k. By a theorem of Néron ([N1]), $A(K)$ is a finitely generated group.

Theorem. (Néron). *There is an abelian variety A_t over k, of the same dimension as A, such that*

$$\text{rank} A_t(k) \geq \text{rank} A(K).$$

The variety A_t is obtained by "specialising" A. This is done as follows: since K is the function field of $\mathbf{P}_n$ and A is defined over K, A comes from some abelian scheme A_U over a non-empty open subset U of $\mathbf{P}_n$. Let $s_1, ..., s_n \in A(K)$ generate $A(K)$. One can view the s_i's as rational sections of the abelian scheme $\pi : A_U \to U$. By replacing U by a smaller open set, one can also assume that the s_i are morphisms [in fact, such a replacement is not necessary: since U is smooth, the s_i can be proved to be everywhere regular on U]. Hence if t is any rational point of U and $s \in A(K)$, $s(t)$ is a well defined rational point of $A_t = \pi^{-1}(t)$. The map $s \mapsto s(t)$ is a homomorphism $\phi_t : A(K) \to A_t(k)$.

The precise form of Néron's theorem is:

Theorem. *The set of $t \in U(k)$ for which ϕ_t is not injective is thin.*

Therefore outside a thin set, rank $A_t(k) \geq$ rank $A(K)$.

We shall use the following criterion on abelian groups.

Criterion. *Let $\phi : M \to N$ be a homomorphism of abelian groups, and n an integer ≥ 2. We assume that,*

1) *M is finitely generated,*
2) *$M/nM \to N/nN$ is injective,*
3) *ϕ is injective on the torsion group of M,*
4) *ϕ defines an isomorphism of $M_n = \{x \in M, nx = 0\}$ onto $N_n = \{x \in N, nx = 0\}$.*

Then ϕ is injective.

Proof.

The kernel I of ϕ is of finite type from (1), and torsion free, from (3). We shall show that $I = nI$, which implies $I = 0$.

Let $x \in I$. We have $\phi(x) = 0 \in nN$, therefore, from (2), $x \in nM$. We write $x = ny$, with $y \in M$. Then $n\phi(y) = \phi(x) = 0$, that is to say $\phi(y) \in N_n$. From (4), there is $z \in M_n$ with $\phi(z) = \phi(y)$. Since $nz = 0$, we have $x = n(y - z)$. But $y - z \in I$, therefore $x \in nI$.

Proof of Néron's theorem.

We use the criterion for $\phi_y : A(K) \to A_t(k)$, with an arbitrary choice of $n \geq 2$. Condition (1) is known. Condition (3) follows from a general fact on abelian schemes: if s is a section of order exactly n, $s(t)$ is also of order n provided that the residue characteristic at t does not divide n - but here we are in characteristic 0. (The n-division points give a sub-scheme A_n étale over U. A section of this scheme which is non-zero at the generic point is non-zero elsewhere.)

To check condition (4), we have to prove that there are no more n-division points in A_t than in A. Take the subscheme A_n which is the kernel of multiplication by $n : A \to A$. As it is étale over the base U, we can decompose it into a disjoint sum of irreducible subschemes

$$A_n = \coprod B_\alpha.$$

Let d_i be the degree of the projection $B_i \to U$, and let I (resp. J) be the set of i's with $d_i = 1$ (resp. $d_i \geq 2$). If t is chosen outside the thin set $\bigcup_{i \in J} \pi(B_i(k))$, we have

$$|A_{t,n}(k)| = |I| = |A_n(K)|,$$

as desired.

To check condition (2), we have to show that, for t outside a suitable thin set, the map

$$A(K)/nA(K) \to A_t(k)/nA_t(k)$$

is injective. It is enough to prove that, for every $\sigma \in A(K)$, $\sigma \notin nA(K)$, there is a thin set Ω_σ such that $\sigma(t) \notin nA_t(k)$ for all $t \notin \Omega_\sigma$. (We construct this for a finite number of σ: some set of representatives of the classes modulo n).

The image of $\sigma : U \to A_U$ is a subvariety of the scheme A_U; let $V_{n,\sigma}$ be the inverse image by the morphism $n : A_U \to A_U$. Then $V_{n,\sigma}$ is a covering of U, and the hypothesis $\sigma \notin nA(K)$ translates to $V_{n,\sigma}$ having no rational section. We then take Ω_σ to be the image of the rational points of $V_{n,\sigma}(k)$. This concludes the proof.

Remarks.
1) The set of $t \in U(k)$ such that $A(K) \to A_t(k)$ is not injective is in fact finite (and not merely thin). This result is due to Silverman, see e.g. [L6], Chap. 12, §2. The proof is a generalisation of that of Manin-Demjanenko given in §5.2.
2) Néron's theorem can be extended to algebraic groups which are extensions of abelian varieties by groups of multiplicative type. More precisely, if G is such a group, defined over K, and Γ is a finitely generated subgroup of $G(K)$, then Γ "specialises injectively" in $G_t(k)$ for all t outside a thin set. [Let $\bar{\Gamma}$ be the set of $\gamma \in G(K)$ such that there exists $m \geq 1$ with $m\gamma \in \Gamma$. An elementary argument shows that $\bar{\Gamma}$ is finitely generated. One then applies to $\bar{\Gamma}$ the same proof as above for $A(K)$.]

This applies in particular to the multiplicative group $\mathbf{G}_m$ (but not to the additive group $\mathbf{G}_a$).
3) Very little is known about specialising non-commutative groups (even matrix groups).

For instance, let

$$A = \begin{pmatrix} 1 & 1 \\ 0 & 1 \end{pmatrix} \quad \text{and} \quad B_T = \begin{pmatrix} 1 & 1 \\ T & 0 \end{pmatrix}.$$

Over $\mathbf{Q}(T)$, A and B_T generate a free group Γ of rank 2; if one specialises T to $t \in \bar{\mathbf{Q}}$ one gets a group Γ_t with a surjective homomorphism $\phi_t : \Gamma \to \Gamma_t$. An explicit description of the t's for which ϕ_t is injective is not known. For partial results see e.g. [LU].

11.2. Elliptic curves of rank ≥ 9 over $\mathbf{Q}$.

Here now is a construction of an elliptic curve over $\mathbf{Q}$ of rank ≥ 9. We consider 9 points $(P_1, ..., P_9)$ of the projective plane. We suppose that these points are "generically independent": if P_i has projective coordinates $(X_i, Y_i, 1)$, we assume that $X_1, ..., X_9, Y_1, ..., Y_9$ are independent indeterminates. Then $K = \mathbf{Q}(P_1, ..., P_9)$ is a purely transcendental extension of $\mathbf{Q}$ of degree 18. There is a unique cubic curve C defined over K passing through $P_1, ..., P_9$, and this cubic is non-singular. There are 9 obvious points on this cubic, and an obvious divisor class of degree 3 given by a straight line section. We denote such a section by D. (If P is the 3rd point of C on the line P_1P_2, we have $D \sim P_1 + P_2 + P$.) We now show that

$$\{\sum_{i=1}^{9} n_i P_i + mD; \sum_{i=1}^{9} n_i + 3m = 0\}, \text{ in } \mathrm{Pic}(C),$$

has rank 9, that is to say that the P_i are linearly independent modulo the section by a line.

To do so, let k_C be the field of definition of C in $\mathbf{P}_2$ (i.e. the extension of $\mathbf{Q}$ generated by the nine ratios of the coefficients of an equation of C); put $m = \text{tr.deg}_{\mathbf{Q}} k_C$ and $n = \text{tr.deg}_{k_C} \mathbf{Q}(P_1, ..., P_9)$. We have $m \leq 9$, $n \leq 9$ and $m + n = 18$, hence $m = n = 9$. This shows that C is "the generic" cubic, and that the P_i's are generically independent over k_C. Hence, any linear equivalence $\sum_{i=1}^{9} m_i P_i + mD \sim 0$ implies $\sum_{i=1}^{9} m_i x_i + mD \sim 0$ for all $x_i \in C$, which is possible only if $m_1 = ... = m_9 = m = 0$.

Now it is clear from Néron's theorem (§11.1) that if we specialise $P_1, ..., P_9$ outside a thin set we obtain a curve of rank ≥ 9.

We give two complements.

1) *Effective construction of such curves.*

This construction can be made in an elementary way, without using Néron's general theorem.

Let $S = \{7, 11, 19, 23, 29, 31\}$. For each $p \in S$, we select an elliptic curve E_p over $\mathbf{F}_p$, in Weierstrass form and nine distinct rational points $P_1(p), ..., P_9(p)$ on E_p with the following properties:

(a) $P_i(p)$ is not 0 (i.e. is not infinity);

(b) $\sum_{i=1}^{9} P_i(p) \neq 0$, i.e. E_p is the only cubic of $\mathbf{P}_2$ going through the $P_i(p)$;

(c) E_7 has no rational point of order 2;

(d) $P_9(11)$ is not divisible by 2 in $E_{11}(\mathbf{F}_{11})$;

(e) for $p = 19, 23, 29, 31$, $E_p(\mathbf{F}_p)$ has a subgroup of type (2,2), and seven of the $P_i(p)$ are divisible by 2, the other two being generators of the 2-component of $E_p(\mathbf{F}_p)$, these two being P_1, P_2 for $p = 19$; P_3, P_4 for $p = 23$; P_5, P_6 for $p = 29$ and P_7, P_8 for $p = 31$.

(For the actual construction of the E_p's, see below).

One now chooses 9 integral points $P_1, ..., P_9$ in affine 2-space with $P_i \equiv P_i(p) \pmod p$ for $i = 1, ..., 9$, $p \in S$. Let E be the cubic through the P_i (which is unique, because of (b)), and let us show that $P_1, ..., P_9$ are independent on E. Indeed suppose there is a relation $\sum m_i P_i \equiv 0$ mod D and let us show by induction on $\sup |m_i|$ that all the m_i are zero. By reduction mod 19, we see that $m_1 P_1 + m_2 P_2$ is divisible by 2 in $E_{19}(\mathbf{F}_{19})$, hence $m_1 \equiv m_2 \equiv 0$ mod 2. Using similarly reduction mod p for $p = 23, 29, 31, 11$, we see that $m_3 \equiv ... \equiv m_9 \equiv 0$ mod 2. Hence $2 \sum \frac{m_i}{2} P_i \sim 0$. But since $E_7(\mathbf{F}_7)$ has no points of order 2, the same is true for $\text{Pic}(E)$. Hence $\sum \frac{m_i}{2} P_i \sim 0$, and the induction hypothesis shows that all m_i are 0.

It remains to see that (a),...,(e) can actually be met.

From Hasse's theorem, the number N_p of points of E_p over $\mathbf{F}_p$ satisfies

$$p+1-2\sqrt{p} \leq N_p \leq p+1+2\sqrt{p},$$

and conversely for every integer N_p satisfying these inequalities there is a curve E_p over $\mathbf{F}_p$ with exactly N_p points.
a) For $p=7$, we take $N_p = 11$. As $11 \leq 1+7+2\sqrt{7}$. there is an elliptic curve E_7 over $\mathbf{F}_7$ having 11 points. We take any 9 of the 10 non-zero points (their sum is the negative of the 10th, therefore $\neq 0$). As the number of points is odd, there is no point of order 2.

Explicitly we can choose $y^2 = x^3 + 2x - 1$, the 11 points being

$x=$	1	2	3	4	5	∞
$y=$	± 3	± 2	± 2	± 1	± 1	∞

b) We take a cubic modulo 11 having 16 points, whence a subgroup of order 8. We take for $P_1(11), ..., P_8(11)$ the 8 distinct points of this subgroup, and $P_9(11)$ a point outside.

For example, $y^2 = x^3 + 4x + 6$ with 16 points

$x=$	1	2	3	4	6	7	8	9	10	∞
$y=$	0	0	± 1	± 3	± 2	± 5	0	± 1	± 1	∞

c) For $p = 19, 23, 29$ and 31, we have $p+1-2\sqrt{p} \leq 28 \leq p+1+2\sqrt{p}$ therefore we can choose a cubic modulo p having 28 points. Such a cubic has a subgroup of order 4. If it is not cyclic, it has type $(2,2)$. If it is cyclic, it contains a unique subgroup of order 2; dividing by this subgroup we obtain a cubic whose points of order 2 can all be shown to be rational over $\mathbf{F}_p$. In the 2 cases one obtains a cubic with a subgroup of type $(2,2)$ and a subgroup Γ of order 7, with $E(\mathbf{F}_p)/\Gamma \cong \mathbf{Z}/2\mathbf{Z} \times \mathbf{Z}/2\mathbf{Z}$. For $p = 19$ we take p_1, p_2 as a basis of $\mathbf{Z}/2\mathbf{Z} \times \mathbf{Z}/2\mathbf{Z}$, and $P_3, ..., P_9$ are the 7 points of Γ. In the same way for $p = 23$ (with P_3, P_4), $p = 29$ (with P_5, P_6) and $p = 31$ (with P_7, P_8).

In the literature, there is an entirely explicit construction [Na].

2) *In place of 9 independent generic points, we fix 8 points and specialise only the 9th.*

Let $P_1, ..., P_8$ be points of $\mathbf{P}_2(\mathbf{Q})$, pairwise distinct, such that no 3 lie on a straight line and no 6 lie on a conic.

The cubics passing through $P_1, ..., P_8$ form a 1-parameter family ("linear pencil"), that is to say the 8 conditions given by $P_1, ..., P_8$ are independent (see [GH, p.481]). These cubics pass through a 9th fixed point P_0. We assume that P_0 is distinct from $P_1, ..., P_8$.

If we blow up the projective plane $\mathbf{P}_2$ at $P_1, ..., P_6$ we obtain a cubic surface S of $\mathbf{P}_3$ in which the P_i are blown up to lines D_i, $(i = 1, ..., 6)$, and the plane cubics passing through $P_1, ..., P_6$ become plane sections of S; those we consider correspond to planes passing through P_7 and P_8, and P_0 is the third point of intersection of the line P_7P_8 with the surface S.

Let P_9 be a generic point of the plane. We shall show that $P_1, ..., P_9$ *are independent modulo* D *on the elliptic curve* C *going through* $P_0, P_1, ..., P_9$.

We denote by Σ the surface obtained by blowing up the points $P_0, ..., P_8$. Let L_i, $i = 0, ..., 8$ be the lines of Σ corresponding to the P_i; let M be a line of $\mathbf{P}_2$ not passing through the P_i and let $\tilde{M}$ be the corresponding curve of Σ. The elementary theory of blowing up (cf. [H], Chapter 5 §3) shows that the Picard group of Σ (which is here equal to the Néron-Severi group) is the free abelian group with basis $\{L_0, L_1, ..., L_8, \tilde{M}\}$ (thus of rank 10).

The linear system on $\mathbf{P}_2$, defined by the cubics passing through $P_0, ..., P_8$, defines a morphism $\pi : \Sigma \rightarrow \mathbf{P}_1$; the fibres of π are the strict transforms of the cubics passing through the P_i; the hypotheses on $P_0, ..., P_8$ insure that these cubics are irreducible. If T is a generic point of $\mathbf{P}_1$ and $C_T = \pi^{-1}(T)$ is the generic fibre at T, consider the restriction homomorphism

$$\mathrm{Pic}(\Sigma) \rightarrow \mathrm{Pic}(C_T).$$

We have an obvious element in its kernel: the class of a special fibre (note that all the fibres are linearly equivalent). On the basis $\{L_0, ..., L_8, \tilde{M}\}$ this class is that of $3\tilde{M} - \sum_{i=0}^{8} L_i$ [indeed, this is the strict transform of a cubic pasing through $P_0, ..., P_8$; the total transform would be $\sim 3\tilde{M}$, but in the strict transform, the lines L_i due to the blowing up are removed].

As the fibres (even special fibres) are irreducible, one knows that the kernel of $\mathrm{Pic}(\Sigma) \rightarrow \mathrm{Pic}(C_T)$ is generated by the class of a fibre, i.e. here by $3\tilde{M} - \sum_{i=0}^{8} L_i$.

We conclude from this that the classes of $\tilde{M}$ and the L_i $(i = 1, ..., 8)$ in $\mathrm{Pic}(C_T)$ are linearly independent over $\mathbf{Z}$. The interpretation of these classes is clear: that of $\tilde{M}$ is the linear section (denoted by D) of the cubic C_T and that of L_i is P_i $(1 \leq i \leq 8)$ modulo the multiples of D.

Now let P_9 be generic in $\mathbf{P}_2$ and take the cubic passing through P_9 (and $P_0, ..., P_8$). This is the same thing as taking a generic C_T and a generic point P_9 on C_T. If we have

$$\sum_{i=1}^{9} n_i P_i \sim mD, \ (n_i \in \mathbf{Z}, m \in \mathbf{Z}),$$

then, as P_9 is generic and $\sum_{i=1}^{8} n_i P_i - mD$ is fixed, we have $n_9 = 0$. As $P_1, ..., P_8$ are linearly independent modulo D we have $n_i = 0$ $(1 \le i \le 8)$ and $m = 0$, therefore $P_1, ..., P_9$ are linearly independent as claimed.

By Néron's theorem, this shows that P_9 can be specialised so that the rank of the corresponding curve is ≥ 9.

Exercise. Show that the construction in this section (resp. in the next) gives an infinite number of curves over $\mathbf{Q}$, with distinct j invariants, of rank ≥ 9 (resp. ≥ 10).

11.3. Elliptic curves of rank ≥ 10 over Q.

We start again from a pencil L of cubics passing through 8 general points $P_1, ..., P_8$. We select 2 of these points, say P_1, P_2, and lines D_1, D_2 passing through P_1 and P_2, respectively (and not passing through any other point P_i). Each cubic $C \in L$ cuts D_1 in P_1 and 2 other points; this defines an involution on D_1, and similarly for D_2. These involutions will be denoted $R \mapsto R'$, resp. $S \mapsto S'$.

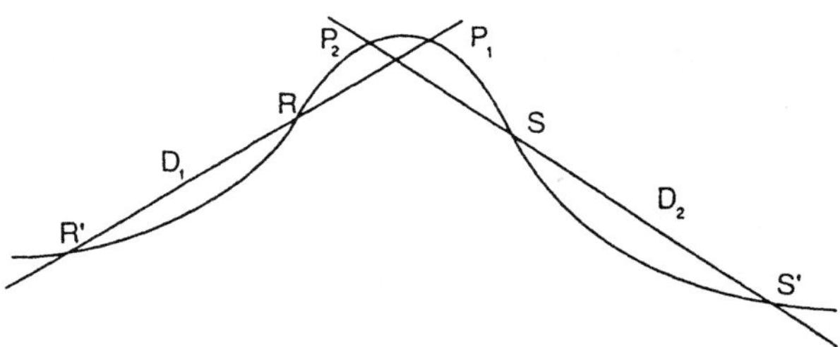

Let now Γ be the (strict) graph of the correspondence on $D_1 \times D_2$ defined by L: a pair $(R, S) \in D_1 \times D_2$ belongs to Γ if there exists $C \in L$ with $C.D_1 = P_1 + R + R'$, $C.D_2 = P_2 + S + S'$.

We have a diagram:

$$\begin{array}{ccccc} & & \Gamma & & \\ & \swarrow & & \searrow & \\ D_1 & & & & D_2 \\ & \searrow & & \swarrow & \\ & & \mathbf{P}_1 & & \end{array}$$

and Γ is the fibre product of D_1 and D_2 over L, hence a (ramified) covering of L of type $(2, 2)$.

Since D_1 is a genus zero covering of degree 2 of L, it is ramified in two points α_1, β_1 of L (which correspond to cubics in L tangent to D_1). Similarly, D_2 is ramified in two points α_2, β_2 of L.

There are now three possibilities:

1) $\{\alpha_1, \beta_1\}$ and $\{\alpha_2, \beta_2\}$ are disjoint. In that case Γ has genus 1 (and is non-singular).
2) $\{\alpha_1, \beta_1\}$ and $\{\alpha_2, \beta_2\}$ have exactly one point in common, say $\alpha_1 = \alpha_2$. In that case Γ has genus 0 (and has one double point, above α_1, which corresponds to a cubic $C \in L$ tangent to both D_1 and D_2).
3) $\{\alpha_1, \beta_1\} = \{\alpha_2, \beta_2\}$. In that case Γ decomposes into 2 curves of genus 0 (over a suitable extension of the ground field).

The favourable case is case 2). Indeed, Γ can then be chosen to reparametrise the pencil L: to $(R, S) \in \Gamma$ we associate the corresponding cubic $C_{R,S}$. On the generic member of this family, we now have 10 points: $P_1, ..., P_8, R, S$. If these can be proved to be independent (modulo the section D by a line, as in §11.2), Néron's theorem will give us elliptic curves of rank ≥ 10 over $\mathbf{Q}$.

In résumé, there are two conditions to be checked:

a) that case 2) occurs;
b) that $P_1, ..., P_8, R, S$ are independent on the generic $C_{R,S}$.

We now show how this can be done. The construction is essentially that of Néron [N1] except for the use of generic data instead of rational data.

Details of the construction.

We start with the generic independent points $Q_1, Q_2, P_3, ..., P_8, X$ of the plane $\mathbf{P}_2$. This gives a purely transcendental extension of $\mathbf{Q}$ of degree 18. We denote by C_1 the cubic passing through these 9 points. We have already seen that the field of definition $\mathbf{Q}(C_1)$ of C_1 is a purely transcendental extension of $\mathbf{Q}$ of transcendence degree 9, and that $Q_1, Q_2, P_3, ..., P_8, X$ are generic independent points on C_1 with respect to $\mathbf{Q}(C_1)$. We consider the pencil L of cubics through $Q_1, Q_2, P_3, ..., P_8$.

We choose a line D_1 (among the 4 possessing this property) passing through Q_1 and tangent to C_1 at a point P_1 with $P_1 \neq Q_1$. The field $\mathbf{Q}(C_1, P_1)$ is a finite extension (of degree 4) of $\mathbf{Q}(C_1, Q_1)$). We choose in the same way a line D_2 passing through Q_2 and tangent to C_1 at P_2 with $P_2 \neq Q_2$. The transcendence degree over $\mathbf{Q}$ of the field $\mathbf{Q}(P_1, ..., P_8, X)$ is then 18, therefore $P_1, ..., P_8, X$ are generic independent points of $\mathbf{P}_2$, and we can view C_1 as the unique cubic passing through $P_1, ..., P_8, X$.

Lemma. *The curve Γ associated to D_1, D_2 and the pencil L of cubics passing through $Q_1, Q_2, P_3, ..., P_8$ has type 2.*

Proof.
We do not have case 1 since the cubic C_1 is tangent to both D_1 and D_2. We must check there is no other, to avoid case 3.

We assume that there is a cubic C' different from C_1 passing through $Q_1, Q_2, P_3, ..., P_8$ tangent to D_1 at P_1' and to D_2 at P_2'.

Let $K = \mathbf{Q}(C_1, P_1, P_2)$. The points Q_1 and Q_2 are rational over K, therefore D_1 and D_2 also. The transcendence degree of $K(C')$ over K is at most 5. Indeed the point P_1' belongs to D_1, therefore depends on one parameter; the same is true for P_2' and the cubics passing through Q_1, Q_2, tangent at P_1', P_2' to D_1, D_2, respectively, depend on 3 parameters, except when P_1' and P_2' are equal to $D_1 \cap D_2$ in which case there are 4 parameters.

On the other hand $\mathbf{Q}(C_1, P_1, ..., P_8) = K(P_3, ..., P_8)$ has transcendence degree 17 over $\mathbf{Q}$, thus 6 over K. But $P_3, ..., P_8$ are in the intersection of C_1 and C', thus are algebraic over $K(C')$, whence a contradiction.

We have seen that $P_1, ..., P_8$ are generic points of $\mathbf{P}_2$, thus the field $\mathbf{Q}(P_1, ..., P_8)$ is a purely transcendental extension of $\mathbf{Q}$ of degree 16, and $E = \mathbf{Q}(P_1, ..., P_8, C_1)$ is a purely transcendental extension of $\mathbf{Q}$ of degree 17.

Let Γ be the curve in $D_1 \times D_2$ defined as above with respect to the pencil L. Since we have type 2, Γ is irreducible and its genus is 0. Moreover Γ contains the point (R, R), where R is the point of intersection of D_1 and D_2; this point is a simple point and is rational over E. Hence Γ *is birational to* $\mathbf{P}_1$.

Let (R, S) be a generic point of Γ with respect to E. The field $F = E(R, S)$ is the function field of Γ; it is a purely transcendental extension of $\mathbf{Q}$ of degree 18. Let C be the cubic passing through $R, S, Q_1, Q_2, P_3, ..., P_8$. As C is defined over a purely transcendental extension of $\mathbf{Q}$, it only remains to prove the following lemma.

Lemma. *The 10 points $R, S, Q_1, Q_2, P_3, ..., P_8$ are independent on C modulo a line section.*

Consider a relation

$$mR + nS + q_1Q_1 + q_2Q_2 + \sum_{i=3}^{8} p_iP_i \sim 0 \quad (\text{mod } D),$$

with $m, n, q_1, q_2, p_i \in \mathbf{Z}$.

Let R' be the third point of intersection of D_1 with C. The point (R', S), the image of (R, S) by an involution, is a generic point of V. Therefore

$$mR' + nS + q_1Q_1 + q_2Q_2 + \sum_{i=3}^{8} p_iP_i \sim 0 \pmod{D}.$$

But R, R', Q_1 are collinear:

$$R + R' + Q_1 \sim 0 \pmod{D}.$$

Whence

$$2nS + (2q_1 - m)Q_1 + 2q_2Q_2 + 2\sum_{i=3}^{8} p_iP_i \sim 0 \pmod{D}.$$

But $S, Q_1, Q_2, P_3, ..., P_8$ are generic independent points of $\mathbf{Q}$. [Indeed, the field $K = \mathbf{Q}(Q_1, Q_2, P_3, ..., P_8)$ has transcendence degree 16 by assumption, and one easily checks that $\text{tr.deg}_K K(D_2) = 1$ and $\text{tr.deg}_{K(D_2)} K(D_2, S) = 1$.] This implies that $S, Q_1, Q_2, P_3, ..., P_8$ are independent (mod D), hence:

$$n = 0, m = 2q_1, q_2 = 0, p_3 = ... = p_8 = 0.$$

A similar argument, with R and S permuted, gives

$$m = 0, n = 2q_2, q_1 = 0, p_3 = ... = p_8 = 0,$$

showing that all the coefficients are 0.

11.4. Elliptic curves of rank ≥ 11 over $\mathbf{Q}$.

Néron [N2] has sketched a method for constructing elliptic curves of rank ≥ 11 over $\mathbf{Q}$. One starts with a pencil L of elliptic curves through points $P_1, ..., P_8$ and lines D_1, D_2, D_3 through P_1, P_2, P_3. As before, the correspondence between D_1, D_2, D_3 defined by L gives a curve $\Gamma \subset D_1 \times D_2 \times D_3$. If one has

a) D_1, D_2, D_3 are concurrrent,

b) there is a cubic L tangent to D_1, D_2 and D_3,

then Γ is (in general) of genus 1. If moreover Γ has infinitely many rational points (R_1, R_2, R_3), then every such point gives a cubic with 11 "obvious" points $(R_1, R_2, R_3, P_1, ..., P_8)$, which one can hope to be independent (in general). Néron only gives hints about all this, but a detailed account can be found in [Fr]; see also [To].

Questions.
1) Can one construct elliptic curves over **Q** of arbitrarily high rank?

The best that Néron's method gives is the existence of infinitely many distinct elliptic curves with rank ≥ 11. Nevertheless, explicit numerical examples (constructed by Mestre [Me]) are known of curves of higher rank; the present record being ≥ 14 (1987). Still higher ranks can probably be obtained by the same method.

2) Is there a bound for the rank of an elliptic curve over $\mathbf{Q}(T)$ with non-constant j-invariant?

3) A different question is the variation with the ground field of the rank of a fixed elliptic curve. For instance, Honda has conjectured ([Ho], misprinted on p.98) that for an abelian variety A over a number field k there is a constant $C = C(A, k)$ such that

$$\text{rank } A(k') \leq C[k' : k],$$

for every finite extension k' of k. This is a strong conjecture which would imply that all the quadratic twists of a given elliptic curve have bounded rank.

12. THE LARGE SIEVE

Let A be a thin set in $\mathbf{Z}^n$, and A_N the intersection of A with the ball of diameter N centred at the origin. When $n = 1$ we have seen in §9.7 (as a consequence of Siegel's theorem) that $|A_N| = O(N^{1/2})$ when $N \to \infty$. To prove a similar result when $n \geq 2$ one needs a different method, based on the large sieve inequality (cf. [Co]).

The present chapter contains a proof of this inequality (after [Bom1] and [Hu]). Applications to thin sets are given in the next chapter.

12.1. Statement of the main theorem.

Let K be a number field, $d = [K : \mathbf{Q}]$, O_K the ring of integers of K, and Λ a torsion free O_K-module of finite type with rank n over O_K. We select a norm $|.|$ on $\Lambda_{\mathbf{R}} = \mathbf{R} \otimes \Lambda$.

[A typical example is $K = \mathbf{Q}$, $O_K = \mathbf{Z}$, $\Lambda = \mathbf{Z}^n$, and for $x = (x_1, ..., x_n) \in \mathbf{Z}^n$ we take $|x| = \sup |x_i|$.]

We consider a subset X of Λ. We impose bounds on the the "size" of X both from the archimedian and non-archimedian viewpoints. More precisely, let N and Q be real numbers, $N \geq 1$, $Q > 0$; for every prime ideal $\wp$ of O_K, let $\omega_\wp$ be a real number in $[0, 1]$. We assume that:

1) The set X is contained in a ball of diameter N: there is $x_0 \in \Lambda_{\mathbf{R}}$ such that $|x - x_0| < N/2$ for all $x \in X$.

2) For every $\wp$ with $\mathbf{N}\wp \leq Q$, the image $X_\wp$ of X in $\Lambda/\Lambda\wp$ by reduction modulo $\wp$ is such that

$$|X_\wp| \leq (1 - \omega_\wp)|\Lambda/\wp\Lambda|.$$

The assumption (2) means that a proportion $\omega_\wp$ of the classes modulo $\wp$ are missing from $X_\wp$; they have been "sieved out".

Theorem. *There is a constant $c = c(\Lambda, |.|)$ depending only on K, Λ and the norm chosen on $\Lambda_{\mathbf{R}}$, such that for all $N, Q, \omega_\wp$, and all X satisfying (1) and (2), we have*

$$|X| \leq c \sup(N^{nd}, Q^{2n})/L(Q)$$

where

$$L(Q) = \sum_{\underline{a}} \prod_{\wp | \underline{a}} \frac{\omega_\wp}{1 - \omega_\wp},$$

the sum being over all ideals $\underline{a}$ of O_K without square factors and of norm $\leq Q$.

The constant c can be made explicit. For example, in the case $\Lambda = \mathbf{Z}^n$, $|x| = \sup |x_i|$, one can take $c = 2^n$.

Example. Take $K = \mathbf{Q}$, $n = 1$, and X the set of squares in $[1, N]$. One can then take $\omega_\wp$ close to $1/2$, and $L(Q) \asymp Q$; choosing $Q \asymp N^{1/2}$ gives $|X| \ll N^{1/2}$, a not very surprising statement!

The name "large sieve" refers to the size of $\omega_\wp$ (i.e. the size of the missing subset for X modulo $\wp$) e.g. $\omega_\wp$ a constant > 0; this is the relevant case for thin sets. In the *small sieve* $\omega_\wp$ may be as small as $1/\mathbf{N}\wp^n$.

The above theorem is an assertion in harmonic analysis; it expresses properties of functions on a torus. The lattice Λ plays the rôle of the group of characters, in the Pontrjagin sense, of a compact torus $T = T_\Lambda$. If $\lambda \in \Lambda$, let χ_λ be the corresponding character on T_Λ. Associated to a subset X of Λ is the function

$$f_X = \sum_{\lambda \in X} \chi_\lambda$$

where the Fourier coefficients are 1 in X and 0 elsewhere. Hypothesis (1) means that the Fourier coefficients of f_X are zero outside a ball of radius N.

Consider now the reduction of Λ modulo $\wp$. By duality, the quotient group $\Lambda/\wp\Lambda$ is the dual group of $T_\wp$, where $T_\wp$ is the subgroup of T of $\wp$-torsion points. Hypothesis (2) means that when the restriction of f to the $\wp$-torsion points is expanded as a finite Fourier series, only a proportion $\leq 1 - \omega_\wp$ of the Fourier coefficients are non-zero.

We shall need two auxiliary results. The first is about functions on finite abelian groups; the second is about the values of Fourier series on sets of well spaced points on a torus.

12.2. A lemma on finite groups.

Lemma. *Let $G_1, \ldots, G_m$ be finite abelian groups written additively. For $1 \leq i \leq m$, let $\hat{G}_i = \mathrm{Hom}(G_i, \mathbf{C}^*)$ be the dual of G_i, Ω_i a subset of $\hat{G}_i$, and $\omega_i = |\Omega_i|/|G_i|$. Let f be a complex valued function on $G_1 \times \ldots \times G_m$. Assume that f is a linear combination of products $\chi_1 \ldots \chi_m$, with $\chi_i \in \hat{G}_i - \Omega_i$. Then*

$$|f(0)|^2 \leq \Big(\prod_{i=1}^{m} \frac{1-\omega_i}{\omega_i}\Big) \sum_{\substack{x_j \in G_j, x_j \neq 0 \\ \text{for all } j=1,\ldots,m}} |f(x_1, \ldots, x_m)|^2.$$

We say that an element of the product $G_1 \times ... \times G_m$ is *primitive* if no component is zero; we can then write the latter sum of the lemma in the form

$$\sum_{x \in G,\ x \text{ primitive}} |f(x)|^2.$$

Proof.

By induction on m.

a) $m = 1$, $G = G_1$.

The Fourier expansion of f is:

$$f = \sum_{\chi \in \hat{G} - \Omega_1} a_\chi \chi.$$

By the orthogonality formulae, the L^2 norm of f is

$$\sum_\chi |a_\chi|^2 = \frac{1}{|G|} \sum_{x \in G} |f(x)|^2.$$

We have

$$f(0) = \sum_\chi a_\chi$$

and by the Cauchy-Schwarz inequality

$$|f(0)|^2 \leq (\sum_\chi |a_\chi|^2) \sum_{\chi \in \hat{G} - \Omega_1} 1$$

$$\leq |G|(1 - \omega_1) \sum_\chi |a_\chi|^2$$

$$\leq (1 - \omega_1) \sum_{x \in G} |f(x)|^2.$$

Therefore

$$|f(0)|^2 \leq (1 - \omega_1)|f(0)|^2 + (1 - \omega_1) \sum_{x \in G, x \neq 0} |f(x)|^2,$$

hence

$$\omega_1 |f(0)|^2 \leq (1 - \omega_1) \sum_{x \in G, x \neq 0} |f(x)|^2.$$

We finally divide by ω_1 to obtain the result (if $\omega_1 = 0$, the lemma is trivial).
b) *General case.*

For $x_m \in G_m$, we define $f_{x_m}(x_1, ..., x_{m-1}) = f(x_1, ..., x_m)$. Then f_{x_m} is a function on $G_1 \times ... \times G_{m-1}$ to which we can apply the induction hypothesis:

$$\sum_{x_i \neq 0, i=1,...,m-1} |f_{x_m}(x_1, ..., x_{m-1})|^2 \geq (\prod_{i=1}^{m-1} \frac{\omega_i}{1-\omega_i}) |f_{x_m}(0)|^2.$$

Hence we have

$$\sum_{x \in G,\ x \text{ primitive}} |f(x)|^2 \geq (\prod_{i=1}^{m-1} \frac{\omega_i}{1-\omega_i}) \sum_{x_m \in G_m, x_m \neq 0} |f(0, ..., 0, x_m)|^2.$$

We then apply part (a) of the proof to the function $f(0, ..., 0, x_m)$ on G_m:

$$\sum_{x_m \in G_m, x_m \neq 0} |f(0, ..., 0, x_m)|^2 \geq \frac{\omega_m}{1-\omega_m} |f(0)|^2,$$

whence the lemma.

This lemma shows that a function involving few characters does not oscillate much. An extreme case is that of one character (the absolute value is constant). At the other extreme, the Dirac δ-function (which equals 1 at the origin and 0 elsewhere) involves all the characters.

12.3. The Davenport-Halberstam theorem.

For the archimedian setting, the O_K-module structure of Λ plays no rôle. We consider then a free $\mathbf{Z}$-module Λ of rank m. Let $\Lambda' = \mathrm{Hom}(\Lambda, \mathbf{Z})$ be the adjoint (or dual) of Λ; $\Lambda_{\mathbf{R}} = \mathbf{R} \otimes \Lambda$ is a vector space over $\mathbf{R}$ of dimension m, its dual is $\Lambda'_{\mathbf{R}} = \mathbf{R} \otimes \Lambda'$, and Λ' is a lattice in $\Lambda'_{\mathbf{R}}$. Let T_Λ be the quotient torus $\Lambda'_{\mathbf{R}}/\Lambda'$; its dual is Λ; if $\lambda \in \Lambda$ we denote by χ_λ the corresponding character of T_Λ:

$$\chi_\lambda(x) = \exp\{2\pi i < \lambda, x >\}, x \in T_\Lambda = \Lambda'_{\mathbf{R}}/\Lambda'.$$

Theorem. (Davenport-Halberstam, see [Bom1],[Hu],[Mo]). *Let Λ be a free abelian group of rank m, and let T_Λ be the dual torus. Select norms on $\Lambda_{\mathbf{R}}$ and on $\Lambda'_{\mathbf{R}}$. There is a constant c, depending on the chosen norms, such that if $\delta > 0$, and $x_1, ..., x_r \in T_\Lambda$ are δ-spaced in the sense that*

$$||x_i - x_j|| > \delta, \text{ for all } i \neq j,$$

and if f is a complex function on T_Λ which is a linear combination of characters χ_λ with $\lambda \in \Lambda$ in a ball of diameter $\leq N$, with $N \geq 1$, then

$$(*) \qquad \sum_{i=1}^{r} |f(x_i)|^2 \leq c \sup(N, \delta^{-1})^m ||f||_2^2.$$

Remarks. 1) The norm $x \mapsto ||x||$ on T_Λ is defined by passage to the quotient from the norm on $\Lambda'_{\mathbf{R}}$:

$$||x|| = \inf_{y \mapsto x} ||y||.$$

2) The L^2-norm $||f||_2$ of f relative to the Haar measure dx with volume 1 is:

$$||f||_2^2 = \int_T |f(x)|^2 \, dx = \sum |a_\lambda|^2,$$

where $f = \sum a_\lambda \chi_\lambda$.

12.4. Proof of the Davenport-Halberstam theorem.

It suffices to prove the theorem for a particular absolute value. We shall do this for $\Lambda = \mathbf{Z}^m$, $T = \mathbf{R}^m/\mathbf{Z}^m$ and $|t| = \sup |t_i|$ for $t = (t_1, ..., t_m) \in \mathbf{R}^m$. With this choice, we shall see that $c = 2^m$ suffices.

There are proofs of this theorem based on different principles. We present here the original method, which is well suited to the case of several variables. The methods which were subsequently invented give finer results for $m = 1$.

a) The theorem is true in the case where $\delta > 1/2$.

Indeed, two distinct rational points x_i, x_j of $\mathbf{R}^m/\mathbf{Z}^m$ satisfy $||x_i - x_j|| \leq 1/2$. If $\delta > 1/2$, then $r \leq 1$; but by hypothesis $f = \sum_{\lambda \in X} a_\lambda \chi_\lambda$, where X is contained in a box of diameter $\leq N$, therefore

$$|f(x_1)|^2 \leq (\sum_{\lambda \in X} |a_\lambda|^2)(\sum_{\lambda \in X} 1) \leq (N+1)^m ||f||_2^2 \leq (2N)^m ||f||_2^2.$$

b) We now assume that $\delta \leq 1/2$. Let ϕ be a continuous complex-valued function on $\mathbf{R}^m$ which is zero for $|t| \geq \delta/2$. Extending ϕ by periodicity, we can consider it as a function on the torus T; the coefficient of χ_λ in its Fourier expansion is then

$$c_\lambda(\phi) = \int_{\mathbf{R}^m} \phi(x)\chi_{-\lambda}\, dx = \int_T \phi(x)\chi_{-\lambda}(x)\, dx.$$

The L^2-norm is given by

$$||\phi||_2^2 = \int_{\mathbf{R}^m} |\phi(x)|^2\, dx = \int_T |\phi(x)|^2\, dx.$$

We now assume:
(†) $|c_\lambda(\phi)| \geq 1$ *for all* $\lambda \in X$.

Lemma 1. *We have* $\sum_{i=1}^r |f(x_i)|^2 \leq ||\phi||_2^2 ||f||_2^2$.

Proof. Consider the function

$$g = \sum_{\lambda \in X} \frac{c_\lambda(f)}{c_\lambda(\phi)} \chi_\lambda.$$

Its Fourier coefficients are

$$c_\lambda(g) = \begin{cases} \frac{c_\lambda(f)}{c_\lambda(\phi)} & \text{if } \lambda \in X \\ 0 & \text{if } \lambda \notin X \end{cases}$$

Thus $c_\lambda(f) = c_\lambda(\phi)c_\lambda(g)$, therefore $f = \phi * g$, where $*$ is the convolution product

$$f(x) = \int_T \phi(x-t)g(t)\, dt.$$

Let B_i be the ball of centre x_i and radius $\delta/2$. As $\phi(x-t) = 0$ for $||x-t|| > \delta/2$, we have

$$f(x_i) = \int_{B_i} \phi(x_i - t)g(t)\, dt,$$

therefore

$$|f(x_i)|^2 \leq ||\phi||_2^2 \int_{B_i} |g(t)|^2\, dt.$$

But the B_i are pairwise disjoint, hence we have

$$\sum_{i=1}^{r} |f(x_i)|^2 \leq ||\phi||_2^2 ||g||_2^2.$$

Finally, as $|c_\lambda(g)| \leq |c_\lambda(f)|$ by (†), we have

$$||g||_2^2 = \sum_{\lambda \in X} |c_\lambda(g)|^2 \leq \sum_{\lambda \in X} |c_\lambda(f)|^2 = ||f||_2^2,$$

whence the lemma.

We now construct such functions ϕ. First, we consider the case of one variable.

Lemma 2. *Let Φ be the function on $\mathbf{R}$ defined by*

$$\Phi(x) = \begin{cases} 2\cos \pi x & \text{for } |x| \leq 1/2, \\ 0 & \text{for } |x| > 1/2. \end{cases}$$

Then the Fourier transform $\hat{\Phi}$ of Φ satisfies

$$\hat{\Phi}(y) \geq 1 \text{ if } |y| \leq 1/2,$$

and

$$||\Phi||_2^2 = \int_{-\infty}^{+\infty} |\Phi(x)|^2 \, dx = 2.$$

Proof.
We have

$$\hat{\Phi}(y) = \int_{-1/2}^{1/2} 2\cos(\pi x) e^{-2\pi i x y} \, dx.$$

As Φ is even, $\hat{\Phi}$ is real:

$$\hat{\Phi}(y) = \int_{-1/2}^{1/2} 2\cos(\pi x)\cos(2\pi x y) \, dx = \frac{1}{\pi} \frac{\cos \pi y}{\frac{1}{4} - y^2}.$$

This shows that $\hat{\Phi}$ is decreasing in the interval $[0, 1/2]$. Other hand, we have

$$\hat{\Phi}(1/2) = \int_{-1/2}^{1/2} 2(\cos \pi x)^2 \, dx = 1/2||\Phi||_2^2 = 1.$$

The lemma follows.

In the general case of m variables, we can then construct the function ϕ in the following way. Put $M = \sup(N, \delta^{-1})$. Let $\lambda_0 \in \mathbf{R}^m$ be the centre of a box of diameter $\leq N$ containing X. For $x = (x^{(1)}, ..., x^{(m)}) \in \mathbf{R}^m$, put

$$\phi(x^{(1)}, ..., x^{(m)}) = M^m \chi_{\lambda_0}(x) \prod_{i=1}^{m} \Phi(Mx^{(i)}),$$

where Φ is as above.

.) If $|x| > \delta/2$, we have $\phi = 0$.

Indeed, $|x| > \delta/2$ implies that there is i such that $|x^{(i)}| > \delta/2$, hence $M|x^{(i)}| > M\delta/2 > 1/2$; therefore $\Phi(Mx^{(i)}) = 0$ and $\phi(x) = 0$.

.) The L^2-norm of ϕ is:

$$||\phi||_2^2 = M^{2m} \prod_{i=1}^{m} \int_{-\infty}^{+\infty} |\Phi(Mx^{(i)})|^2 \, dx^{(i)} = 2^m M^m.$$

.) The absolute value of the Fourier transform of ϕ is ≥ 1 in the box with centre λ_0 and diameter N.

As multiplication by χ_{λ_0} just translates the Fourier transform, it suffices to see that the absolute value of the Fourier transform of the function $x \mapsto M^m \prod_{i=1}^{m} \Phi(Mx^{(i)})$ is ≥ 1 on the box with centre 0 and diameter N. The value of this Fourier transform at the point $(y^{(1)}, ..., y^{(m)})$ is

$$M^m \prod_{i=1}^{m} \int_{-\infty}^{+\infty} \Phi(Mx^{(i)}) \exp(-2\pi i (Mx^{(i)})(y^{(i)}/M)) \, dx^{(i)}$$

$$= \prod_{i=1}^{m} \hat{\Phi}(y^{(i)}/M).$$

If $|y^{(i)}| \leq N/2$, then $|y^{(i)}/M| \leq 1/2$, therefore $|\prod_{i=1}^{m} \hat{\Phi}(y^{(i)}/M)| \geq 1$.

The Davenport-Halberstam theorem now follows from lemma 1 and the formula

$$||\phi||_2^2 = 2^m M^m = 2^m \sup(N, \delta^{-1})^m.$$

Commentary on the constants.

We restrict ourselves to the case $m = 1$ (Fourier series in one variable) which is the only case that has been studied in detail.

We consider two numbers $N \geq 1$ and $\delta \leq 1$, and we seek the best possible constant $\Delta = \Delta(N, \delta)$ for which

$$\sum_{i=1}^{r} |f(x_i)|^2 \leq \Delta(N, \delta) ||f||_2^2$$

given that the x_i are δ-spaced and f is a trigonometric polynomial of length $\leq N$:

$$f(x) = \sum_{j=j_0}^{j_0+N} c_j e^{-2\pi i j x}.$$

We have seen that:

$$\Delta(N, \delta) \leq 2 \sup(N, \delta^{-1}).$$

For arbitrary N and δ, the exact value of Δ is not known, but one has good bounds for it.

Theorem. (Selberg, see [Mo]). *We have* $\Delta \leq N + \delta^{-1}$.

This can be proved by the preceding method, if $N\delta \leq 1$, but the interesting case for some applications is $N\delta > 1$, for which we refer to [Mo].

This Selberg bound is optimal for some values of N and δ. For example, take $x_1, ..., x_r$ to be the M-torsion points on $T = \mathbf{R}/\mathbf{Z}$, with $r = M$. We choose $f(x) = \theta(Mx)$, where

$$\theta(x) = \sum_{n=0}^{L} e^{2\pi i n x}, \text{ for some } L \geq 1.$$

We may take

$$\delta = 1/M, N = ML.$$

We have

$$f(x_j) = \theta(0) = L + 1$$

$$\sum_{j=1}^{r} |f(x_j)|^2 = (L+1)^2 M$$

and

$$||f||_2^2 = L + 1,$$

therefore

$$\sum_{j=1}^{r} |f(x_j)|^2 = (N+\delta^{-1})||f||_2^2$$

in this particular case. Note that $N\delta = L \geq 1$.

Bombieri and Davenport obtained a bound which is better when $N\delta$ is small:

$$\Delta(N,\delta) \leq \delta^{-1}(1+270(N\delta)^3).$$

They also constructed examples with $N\delta$ arbitrarily small and

$$\Delta(N,\delta) \geq \delta^{-1}(1+c(N\delta)^3),$$

for some explicit value of c (they give $c = 1/12$, but this can be improved to $c = 5/8$).

Therefore $\Delta(N,\delta)$ is known reasonably well when $N\delta$ is small. For $N\delta \geq 1$, Selberg's bound remains the most useful, see [Bom1], [Mo].

Exercise. Show that

$$\Delta(N,\delta) \leq \frac{1}{M}\Delta(MN,\delta/M),$$

for any integer $M \geq 1$.

12.5. End of the proof of the main theorem.

First, we show that the torsion points on a torus are well spaced. We go back to the notation of §12.1 and §12.3: K is a number field, $[K:\mathbf{Q}] = d$, Λ is an O_K-module of rank n, $\Lambda \cong \mathbf{Z}^m$ with $m = nd$. The dual torus $T = \Lambda'_{\mathbf{R}}/\Lambda'$ of Λ also has an O_K-module structure. If $\underline{q}$ is a non-zero ideal of O_K, a point $x \in T$ is called a $\underline{q}$-division point if $\underline{q}.x = 0$; such a point has finite order.

Lemma. *There is a constant c_1 (depending on K and the norm on $\Lambda'_{\mathbf{R}}$) such that the $\underline{q}$-division points on T by the ideals $\underline{q}$ with $\mathbf{N}\underline{q} \leq Q$ are δ-spaced with $\delta = cQ^{-2/d}$.*

The O_K-module Λ' is known to be isomorphic to a direct sum of ideals of O_K:

$$\Lambda' = \underline{b}_1 \oplus ... \oplus \underline{b}_n.$$

Then

$$\mathbf{Q} \otimes \Lambda' = K \oplus ... \oplus K,$$

and

$$\Lambda'_{\mathbf{R}} = K_{\mathbf{R}} \oplus ... \oplus K_{\mathbf{R}},$$

where $K_{\mathbf{R}} = \mathbf{R} \otimes K$ is the product of the archimedian completions of K. We define an absolute value on $K_{\mathbf{R}}$ by putting

$$|x| = \sup_{\sigma} |\sigma x| \quad \text{for } x \in K,$$

where σ runs over the embeddings $K \to \mathbf{C}$, and one extends this to $K_{\mathbf{R}}$ by continuity.

For $x = (x^{(1)}, ..., x^{(n)}) \in \Lambda'_{\mathbf{R}}$, we take

$$|x| = \sup_{\sigma, i} |\sigma(x^{(i)})|.$$

For this particular norm, we prove the lemma with

$$c = \inf_i \mathbf{N}(\underline{b}_i)^{1/d}.$$

We take a $\underline{q}$-division point t and a $\underline{q}'$-division point t' with $t \neq t'$, $\mathbf{N}\underline{q} \leq Q$ and $\mathbf{N}\underline{q}' \leq Q$. Then $x = t - t'$ is a $\underline{qq}'$-division point and $\mathbf{N}\underline{qq}' \leq Q^2$. Write $x = (x^{(1)}, ..., x^{(n)})$ where $x^{(i)} \in K$. The condition

$$\underline{qq}'x \subset \Lambda' = \underline{b}_1 \oplus ... \oplus \underline{b}_n$$

means that

$$\underline{qq}'x^{(i)} \subset \underline{b}_i, \quad (1 \leq i \leq n).$$

Let i be such that $x^{(i)} \neq 0$. Then

$$\mathbf{N}(\underline{qq}'x^{(i)}) \geq \mathbf{N}(\underline{b}_i);$$

therefore

$$\prod_{\sigma} |\sigma x^{(i)}| \geq \mathbf{N}(\underline{b}_i)/Q^2, \text{ since } |\mathbf{N}(x^{(i)})| = \prod_{\sigma} |\sigma x^{(i)}|,$$

and

$$|x^{(i)}| \geq \mathbf{N}(\underline{b}_i)^{1/d}/Q^{2/d} \geq c_1 Q^{-2/d},$$

whence the lemma.

We return to the hypotheses of the large sieve (§12.1): given $\Lambda, T, Q, N, \{\omega_\wp\}$ the set X is contained in a ball of diameter $N \geq 1$, and its image $X_\wp$ in $\Lambda/\wp\Lambda$ satisfies

$$|X_\wp| \leq (1 - \omega_\wp)[\Lambda : \wp\Lambda],$$

for all prime ideals $\wp$ with $\mathbf{N}\wp \leq Q$.

Take a function $f = \sum_\lambda c_\lambda \chi_\lambda$ on the torus T whose spectrum $\{\lambda \in \Lambda, c_\lambda \neq 0\}$ equals X.

Theorem. *There is a constant* $c = c(\Lambda, |.|)$ *such that for all* $x \in T$, *we have*

$$|f(x)|^2 \le c \sup(N^{nd}, Q^{2n}) ||f||_2^2 / L(Q),$$

where

$$L(Q) = \sum_{\substack{N\underline{q} \le Q \\ \underline{q} \text{ square-free}}} \prod_{\wp | \underline{q}} \frac{\omega_\wp}{1 - \omega_\wp}.$$

It suffices to show this for $x = 0$. Indeed, the translation $x \mapsto x + x_0$ has the effect of multiplying $c_\lambda \chi_\lambda$ by the constant $\chi_\lambda(x_0)$, therefore it does not change the spectrum.

Let T_Q be the set of $\underline{q}$-division points of T for all ideals $\underline{q}$ of O_K with $N\underline{q} \le Q$. We write $T[\underline{q}] = \{t, \underline{q}t = 0\}$. Thus

$$T_Q = \bigcup_{N\underline{q} \le Q} T[\underline{q}].$$

From the above lemma, the set T_Q is δ-spaced, with $\delta = c_1 / Q^{2/d}$. The Davenport-Halberstam theorem (§12.3) then gives

$$\sum_{x \in T_Q} |f(x)|^2 \le c_2 \sup(N^{nd}, \delta^{-nd}) ||f||_2^2$$

$$\le c_3 \sup(N^{nd}, Q^{2n}) ||f||_2^2,$$

with $c_3 = c_2 \sup(1, c_1^{-nd})$.

We say that a $\underline{q}$-division point x is *primitive* if the conditions $\underline{q}' | \underline{q}$ and $\underline{q}' \ne \underline{q}$ imply $\underline{q}' x \ne 0$. Each point $x \in T_Q$ is primitive for a unique $\underline{q}$, namely its annihilator $\{\alpha \in O_K, \alpha x = 0\}$.

We restrict ourselves to ideals $\underline{q}$ without square factors: $\underline{q} = \wp_1 \ldots \wp_h$, where the $\wp_i$ are pairwise distinct prime ideals. In this case

$$T[\underline{q}] = \oplus_{i=1}^h T[\wp_i],$$

and a point x of $T[\underline{q}]$ is primitive if and only if each of its components in the $T[\wp_i]$ is different from 0.

We have

$$\sum_{x \in T_Q} |f(x)|^2 \ge \sum_{\substack{N\underline{q} \le Q \\ \underline{q} \text{ square-free}}} \sum_{\substack{x \in T[\underline{q}] \\ x \text{ primitive}}} |f(x)|^2.$$

We shall now apply the lemma on finite groups of §12.2 to show that

$$\sum_{\text{primitive } x\in T[\underline{q}]} |f(x)|^2 \geq |f(0)|^2 \prod_{\wp|\underline{q}} \frac{\omega_\wp}{1-\omega_\wp}.$$

First note that the dual of $T[\underline{q}]$ is $\Lambda/\underline{q}\Lambda$: the injection

$$T[\underline{q}] \hookrightarrow T$$

is dual to the projection

$$\Lambda \to \Lambda/\underline{q}\Lambda.$$

As the spectrum of f on T is X, it follows that the spectrum of the restriction of f to $T[\underline{q}]$ is contained in the image $X_{\underline{q}}$ of X under the homomorphism $\Lambda \to \Lambda/\underline{q}\Lambda = \prod_{\wp|\underline{q}} \Lambda/\wp\Lambda$. Taking account of the given upper bounds of the $|X_\wp|$, we may then apply §12.2 and obtain the above inequality. By summation over $\underline{q}$ this gives

$$\sum_{x\in T_Q} |f(x)|^2 \geq |f(0)|^2 \sum_{\substack{\mathrm{N}\underline{q}\leq Q \\ \underline{q}\ \text{square-free}}} \prod_{\wp|\underline{q}} \frac{\omega_\wp}{1-\omega_\wp}$$

$$\geq |f(0)|^2 L(Q),$$

which proves the theorem above.

To obtain the large sieve theorem stated in §12.1, we choose $f = \sum_{\lambda\in X} \chi_\lambda$; we have

$$f(0) = |X| = ||f||_2^2,$$

hence the theorem above gives

$$|X|^2 \leq c \sup(N^{nd}, Q^{2n})|X|/L(Q);$$

dividing by $|X|$ gives the result (if $|X| = 0$ the theorem is obviously true).

Example. If $K = \mathbf{Q}$, $\Lambda = \mathbf{Z}^n$, $d = 1$, with the usual norm $\sup |x_i|$, the proof above of the large sieve gives the constant $c = 2^n$:

$$|X| \leq 2^n \sup(N^n, Q^{2n})/L(Q).$$

Remark.

We can give a lower bound for $L(Q)$ in which the sum runs over all ideals (and not only over the square-free ideals $\underline{q}$ of bounded norm); for this we define

$$\omega_{\underline{q}} = \prod_{\wp | \underline{q}} \omega_{\wp}^{n(\wp)} \text{ for } \underline{q} = \prod_{\wp} \wp^{n(\wp)}$$

and

$$L'(Q) = \sum_{\mathrm{N}\underline{q} \leq Q} \omega_{\underline{q}}.$$

We have

$$L(Q) \geq L'(Q).$$

Indeed, the radical $\sqrt{\underline{q}}$ of $\underline{q}$ is $\prod_{\wp|\underline{q}} \wp$; as

$$L'(Q) = \sum_{\substack{\mathrm{N}\underline{q} \leq Q \\ \underline{q} \text{ square-free}}} \sum_{\substack{\mathrm{N}\underline{q}' \leq Q \\ \sqrt{\underline{q}'} = \underline{q}}} \omega_{\underline{q}'},$$

it suffices to check

$$\sum_{\sqrt{\underline{q}'} = \underline{q}} \omega_{\underline{q}'} \leq \frac{\omega_{\underline{q}}}{\prod_{\wp|\underline{q}} (1 - \omega_{\wp})}.$$

But if we write $\underline{q}' = \underline{q} \prod_{\wp|\underline{q}} \wp^{m(\wp)}$, with $m(\wp) \geq 0$, we have $\omega_{\underline{q}'} = \omega_{\underline{q}} \prod \omega_{\wp}^{m(\wp)}$ and

$$\sum_{\underline{q}'} \omega_{\underline{q}'} \leq \omega_{\underline{q}} \prod_{\wp|\underline{q}} \sum_{m(\wp)} \omega_{\wp}^{m(\wp)} \leq \omega_{\underline{q}} \prod_{\wp|\underline{q}} \frac{1}{1 - \omega_{\wp}}.$$

13. APPLICATION OF THE LARGE SIEVE TO THIN SETS

13.1. Statement of results.

1) *Affine thin sets.*

Let K be a number field of degree d, and O_K the ring of integers of K. For $x \in O_K$, we put

$$\ulcorner x \urcorner = \max_{\sigma} |\sigma x|$$

where σ runs over the embeddings of K in $\mathbf{C}$.

Let n be an integer ≥ 1. Let $x = (x^{(1)}, ..., x^{(n)}) \in O_K^n$. We put

$$\ulcorner x \urcorner = \max_{1 \leq i \leq n} \ulcorner x^{(i)} \urcorner.$$

Finally let X be an integer ≥ 1. If M is a subset of O_K^n we define

$$M(X) = |\{x \in M; \ulcorner x \urcorner \leq X\}|.$$

For $M = O_K^n$ this number is $\sim X^{nd}$, when $X \rightarrow \infty$.

Assume now that M is thin in $\mathbf{A}_n(K) \cap O_K^n$ cf. §9.1.

Theorem 1. (S. Cohen). *There is $\gamma < 1$ such that*

$$M(X) = O(X^{(n-1/2)d}(\log X)^{\gamma}).$$

Cohen only proved this result with $\gamma = 1$, but made precise the O as a function of the coefficients of equations defining M [Co]. The proof gives a uniform bound for the number of points of M in every box of diameter X. Note that we gain a factor of $X^{-d/2}(\log X)^{\gamma}$ over the trivial bound $O(X^{nd})$: the set of $x = (x_1, ..., x_n)$ of O_K^n with x_1 a square shows that this improvement is essentially optimum.

2) *Number of integral points on affine varieties.*

Let N be an integer ≥ 1, and V an irreducible algebraic variety of dimension n in affine space $\mathbf{A}_N$ over K. Let $V(X)$ be the number of points $x \in V$ with $x^{(i)} \in O_K$, and $\ulcorner x \urcorner \leq X$. We suppose that V is not linear, i.e. that $\deg V \geq 2$.

Theorem 2. *$V(X) = O(X^{(n-1/2)d}(\log X)^\gamma)$ with $\gamma < 1$.*

This result is derived from theorem 1 as follows. Select coordinates of $\mathbf{A}_N$ such that the projection onto the first n coordinates induces a finite morphism $\pi : V \to \mathbf{A}_n$. The image $\pi(V)$ is then a thin set to which we can apply theorem 1. Each fibre of π has a finite bounded number of points, whence theorem 2.

The same reasoning gives the bound $O(X^{nd})$ when $\deg V = 1$.

3) *Thin projective sets.*

On projective space $\mathbf{P}_n$ over the field K, let H be the standard height associated to an hyperplane section (normalised so that H is invariant under finite extension cf. §2.2). Consider a thin subset $M \subset \mathbf{P}_n(K)$, and denote by $M_K(X)$ the number of $x \in M$ with $H(x) \leq X$.

Theorem 3. *We have $M_K(X) = O(X^{(n+1/2)d}(\log X)^\gamma)$ with $\gamma < 1$.*

By Schanuel's theorem cf. §2.5, the number of $x \in \mathbf{P}_n(K)$ with $H(x) \leq X$ is $\sim cX^{(n+1)d}$. We therefore gain a factor of $X^{-d/2}(\log X)^\gamma$ on the trivial bound.

It would be interesting to know if this bound is the best possible. It seems more reasonable to expect a bound of the form $O(X^{nd}(\log X)^c)$.

The proof that theorem 1 implies theorem 3 will be given in §13.4.

4) *Number of rational points on projective varieties.*

Let V be an irreducible projective variety over K of dimension n, in projective space $\mathbf{P}_N$. We assume that V is not a linear variety. Define $V_K(X) =$ number of $x \in V(K)$ with $H(x) \leq X$.

Theorem 4. *There is $\gamma < 1$ such that*

$$V_K(X) = O(X^{(n+1/2)d}(\log X)^\gamma).$$

Question. Can this bound be improved to $O(X^{nd}(\log X)^c)$ for some c? The factor $(\log X)^c$ is inevitable as the quadric $xy = zt$ in $\mathbf{P}_3$ shows.

Exercise. What about quadrics in n-space, $n \geq 3$?

Theorem 4 is follows from theorem 3 by projecting just as theorem 2 follows from theorem 1.

Properties of thin sets.

While proving theorem 1, we shall obtain an interesting property of the reduction modulo $\wp$ of a thin set.

Let $\wp$ be a prime ideal of O_K. The residue field $k_\wp = O_K/\wp$ is finite, and the map of reduction modulo $\wp$ is everywhere defined on projective space:

$$\mathbf{P}_n(K) \to \mathbf{P}_n(k_\wp).$$

[This is clear from the point of view of schemes. If O_K is principal, take coordinates $(x_1, ..., x_n)$ of $x \in \mathbf{P}_n(K)$ with $x_i \in O_K$ and $x_1, ..., x_n$ coprime, and reduce them modulo $\wp$. In the general case where O_K is not principal, one uses the fact that the local ring at $\wp$ is principal.]
Hence, a subset $M_\wp$ of $\mathbf{P}_n(k_\wp)$ is associated to any given subset M of $\mathbf{P}_n(K)$. If $S = \{\wp_1, ..., \wp_h\}$ is a finite set of prime ideals, let us denote by M_S the image of M in

$$\mathbf{P}_n(k_{\wp_1}) \times ... \times \mathbf{P}_n(k_{\wp_h}) = \mathbf{P}_{n,S}.$$

Theorem. *If M is thin, there is a finite set of prime ideals S such that $M_S \neq \mathbf{P}_{n,S}$.*

Indeed, it is enough to prove this when M is of type 1 or type 2, in the sense of §9.1. For type 1, the assertion is immediate: we have $M_\wp \neq \mathbf{P}_n(k_\wp)$ for almost all $\wp$'s. For type 2, it follows from the corollary to theorem 5 below (§13.2).

Remarks. 1) When $K = \mathbf{Q}$, $n = 1$, the theorem implies that the complement of a thin set contains an arithmetic progression, cf. [Schi].
2) The set S cannot always be taken to have one element, see Exercise.

Exercise. Consider the following thin subsets of $\mathbf{P}_1(K)$:

$$\begin{aligned} M_1 &= \{x \in \mathbf{P}_1(K) | \quad x \text{ is a square}\} \\ M_2 &= \{x \in \mathbf{P}_1(K) | \quad x+1 \text{ is a square}\} \\ M_3 &= \{x \in \mathbf{P}_1(K) | \quad x(x+1) \text{ is a square}\}. \end{aligned}$$

Let $M = M_1 \cup M_2 \cup M_3$. Show that, for every prime ideal $\wp$, the map $M \to \mathbf{P}_1(k_\wp)$ is surjective.

13.2. Proof of theorem 1.

Let $M \subset O_K^n$ be a thin set. An easy redution allows us to assume that $M = O_K^n \cap \pi(Z(K))$, where Z is an irreducible affine variety of dimension

n, and π is a finite morphism of degree ≥ 2, $\pi : Z \to \mathbf{A}_n$. Furthermore, one can replace this by a finite morphism of schemes of finite type over O_K such that

$$M = O_K^n \cap \pi(Z(O_K)).$$

We then say that such a thin set M is primitive.

Let $\wp$ be a prime ideal of O_K. Let $M_\wp$ be the image of M under reduction modulo $\wp$ in $k_\wp \times \ldots \times k_\wp = \mathbf{A}_n(k_\wp)$.

Theorem 5. *There is a finite Galois extension K_π/K and a constant c_π with $0 < c_\pi < 1$ such that if $\wp$ splits completely in K_π then*

$$|M_\wp| \leq c_\pi \mathbf{N}\wp^n + O(\mathbf{N}\wp^{n-1/2}).$$

Corollary. *If $\mathbf{N}\wp$ is large and $\wp$ splits completely in K_π, then $M_\wp \neq \mathbf{A}_n(k_\wp)$.*

The proof will be given in §13.3. We now prove theorem 1, using theorem 5.

We apply the large sieve inequality (§12.1) to M:

$$M(X) \leq c_{K,n} \sup(X^{nd}, Q^{2n})/L(Q),$$

where

$$L(Q) = \sum_{\substack{\mathbf{N}\underline{a} \leq Q \\ \underline{a} \text{ square-free}}} \prod_{\wp | \underline{a}} \frac{\omega_\wp}{1 - \omega_\wp},$$

and

$$1 - \omega_\wp = \frac{|M_\wp|}{\mathbf{N}\wp^n}.$$

The constant $c_{K,n}$ depends only on K and n.

We select $Q = X^{d/2}$. Therefore

$$M(X) \leq c_{K,n} X^{nd}/L(Q), \quad (Q = X^{d/2}),$$

and it only remains to show that

$$L(Q) \gg Q/(\log Q)^\gamma, \quad \gamma < 1.$$

We begin with an easier result.

Lemma 1. *We have $L(Q) \gg Q/\log Q$.*

To see this, we only keep the prime ideals in the sum defining $L(Q)$:

$$L(Q) \geq \sum_{\mathbf{N}\wp \leq Q} \frac{\omega_\wp}{1-\omega_\wp}.$$

This is bounded below by the prime ideals which split (completely) in $K_\wp$; for these by theorem 5 we have $\omega_\wp \geq \epsilon$, with $\epsilon > 0$, for $\mathbf{N}\wp$ sufficiently large; hence:

$$L(Q) \geq \epsilon \sum_{\substack{n_0 \leq \mathbf{N}\wp \leq Q \\ \wp \text{ splits in } K_\pi}} 1.$$

By the prime number theorem, we have

$$\sum_{\substack{\mathbf{N}\wp \leq Q \\ \wp \text{ splits in } K_\pi}} 1 \sim \frac{1}{[K_\pi : K]} \frac{Q}{\log Q},$$

whence lemma 1.

To improve this estimate and get an exponent $\gamma < 1$, we write

$$L(Q) \geq \sum_{\mathbf{N}\underline{a} \leq Q} c_{\underline{a}},$$

where

$$c_{\underline{a}} = \begin{cases} \prod_{\wp | \underline{a}} \frac{\omega_\wp}{1-\omega_\wp} & \text{if } \underline{a} \text{ is squarefree and all } \wp | \underline{a} \text{ split completely in } K_\pi, \\ 0 & \text{otherwise.} \end{cases}$$

We introduce the function

$$f(s) = \sum c_{\underline{a}} \mathbf{N}\underline{a}^{-s}.$$

This is a Dirichlet series, and if we write

$$f(s) = \sum a_n n^{-s},$$

then

$$L(Q) \geq \sum_{n \leq Q} a_n.$$

As $\underline{a} \mapsto c_{\underline{a}}$ is a multiplicative function, we have

$$f(s) = \prod_{\wp}(1 + c_{\wp}\mathrm{N}\wp^{-s})$$

where

$$c_{\wp} = \begin{cases} \frac{\omega_{\wp}}{1-\omega_{\wp}} & \text{if } \wp \text{ splits completely in } K_{\pi}, \\ 0 & \text{otherwise.} \end{cases}$$

If $\wp$ splits completely in $K_{\wp}$, we have from theorem 5 that

$$c_{\wp} = \frac{\mathrm{N}\wp^{n}}{|M_{\wp}}| - 1 = \frac{1}{c_{\pi}} - 1 + O(\mathrm{N}\wp^{-1/2}).$$

Write

$$c = \frac{1}{c_{\pi}} - 1 > 0.$$

Then there are constants $c'_{\wp} > 0$ such that

$$c_{\wp} \geq c'_{\wp} \text{ and } |c - c'_{\wp}| \ll \mathrm{N}\wp^{-1/2},$$

for all $\wp$ which split completely in K_{π}. We introduce the functions

$$f_1(s) = \prod_{\wp \text{ splits in } K_{\pi}} (1 + c'_{\wp}\mathrm{N}\wp^{-s}),$$

and

$$f_2(s) = \prod_{\wp \text{ splits in } K_{\pi}} (1 - \mathrm{N}\wp^{-s})^{-c},$$

for Re $s > 1$. By taking logarithms, one finds that

$$f_1(s) = f_2(s)g(s)$$

for $g(s) = \exp(h(s))$, where h is a Dirichlet series which converges for Re $s > 1/2$.

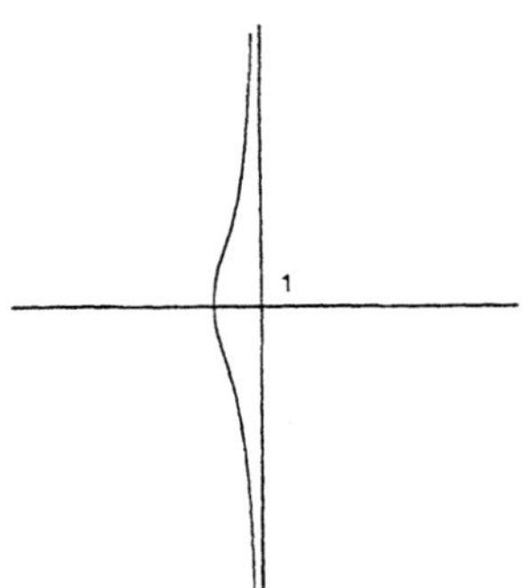

One shows easily that

$$f_2(s) = \zeta_{K_{\pi}}(s)^{\lambda} g_1(s), \text{ with } \lambda = c/[K_{\pi} : K],$$

where $g_1(s) = \exp(h_1(s)))$, h_1 converges for Re $s > 1/2$ and $\zeta_{K_\pi}(s)$ is the zeta function of the number field K_π. Hence

$$f_1(s) = \zeta_{K_\pi}(s)^\lambda g(s) g_1(s)$$

has no zeros or poles except at $s = 1$ in an open set containing Re $s \geq 1$, and

$$f_1(s) \sim c' \frac{1}{(s-1)^\lambda} \quad \text{for } s \text{ in a neighbourhood of } 1.$$

If we had $\lambda = 1$, Ikehara's Tauberian theorem would give for the coefficients $a_{1,n}$ of f_1:

$$\sum_{n \leq Q} a_{1,n} \sim c'Q.$$

For arbitrary λ, an analogous result proved by Delange [De] and Raikov [Ra] gives

$$\sum_{n \leq Q} a_{1,n} \sim c'' \frac{Q}{(\log Q)^{1-\lambda}},$$

with

$$c'' = \frac{c'}{\Gamma(\lambda)}.$$

[This method can be refined using a method of Landau (cf. [Se4]) which gives an asymptotic expansion for $\sum_{n \leq Q} a_{1,n}$.]

This proves theorem 1 with $\gamma = 1 - \lambda$.

13.3. Proof of theorem 5.

Let $\pi : Z \to \mathbf{A}_n$ be as in §13.2. Let $K(Z)$ be the function field of Z, a finite extension of degree $r > 1$ of $K(T_1, ..., T_n)$, and Ω the Galois closure of $K(Z)$ over $K(T_1, ..., T_n)$.

Let K_π be the algebraic closure of K in Ω; this the largest subfield of Ω which is algebraic over K.

[A typical example is $n = 1$, $Z = \mathbf{A}_1$, and $\pi(z) = z^3$.

$$\begin{array}{cc} & \Omega \\ & \uparrow \\ Z & K(\theta) \\ \downarrow \pi & \uparrow \\ \mathbf{A}_1 & K(T) \end{array}$$

The corresponding field extension $K(Z)/K(T)$ is $K(\theta)/K(T)$ with $\theta^3 = T$; hence Ω is the compositum of $K(\theta)$ and K_π where K_π is the, at most quadratic, extension of K obtained by adjoining the cube roots of unity.]

Let Z_Ω be the affine variety which is the normalisation of Z in Ω. Then there is a sufficiently small non-empty open subscheme U of $\mathbf{A}_n$ over which the corresponding open subsets $U_Z \subset Z$, $U_\Omega \subset Z_\Omega$ are finite and étale. Let $G = \mathrm{Gal}(\Omega/K(T))$ and $H = \mathrm{Gal}(\Omega/K(Z))$.

$$\begin{array}{c} U_\Omega \\ \downarrow \\ U_Z \\ \downarrow \\ U \end{array} \qquad H: U_\Omega \to U_Z, \quad G: U_\Omega \to U$$

The field of constants of U_Ω is K_π; in other words, U_Ω is absolutely irreducible over K_π.

Lemma. (Lang-Weil). *If B is an O_K-scheme of finite type whose generic fibre is an absolutely irreducible variety of dimension n over K, then*

$$|B_\wp| = \mathbf{N}\wp^n + O(\mathbf{N}\wp^{n-1/2}).$$

[Here $B_\wp$ means $B(K_\wp)$.]

One possible proof is to fibre the variety by curves and use the Weil estimates on each fibre plus induction on the dimension. This is essentially the method of Lang and Weil, see [LW].

The lemma can also be proved using the formula (Grothendieck - Deligne):

$$|B_\wp| = \sum \pm\lambda_i(\wp)$$

where the $\lambda_i(\wp)$ are the eigenvalues of the Frobenius acting on l-adic cohomology with compact support. Since the reduction of the variety B modulo $\wp$ is absolutely irreducible, for almost all $\wp$, we have $\lambda_i(\wp) = \pm\mathbf{N}\wp^n$ for exactly one i, and for the others $|\lambda_i(\wp)| \leq (\mathbf{N}\wp)^{n-1/2}$. The number of eigenvalues appearing in the formula is uniformly bounded (for an explicit bound in terms of the degrees of defining equations, see [Bom2]).

It remains to show that if $\wp$ splits completely in K_π, the image of $U_{Z,\wp} = U_Z(K_\wp)$ in $U_\wp = U(K_\wp)$ has $\leq c\mathbf{N}\wp^n + O(\mathbf{N}\wp^{n-1/2})$ number of elements, with $c < 1$. We shall even obtain $c \leq 1 - 1/r!$, where $r = [U_Z : U]$, as above.

We change nothing by replacing K by K_π. We may therefore suppose that $K = K_\pi$; then U_Ω is absolutely irreducible over K, and the lemma above gives

$$|U_{\Omega,\wp}| = \mathbf{N}\wp^n + O(\mathbf{N}\wp^{n-1/2}).$$

$$H\left\{\begin{array}{c} U_{\Omega,\wp} \\ \downarrow \\ U_{Z,\wp} \end{array}\right. \hspace{-1em}\left.\begin{array}{c} \\ \\ \\ \downarrow \\ U_\wp \end{array}\right\} G$$

As U_Ω and U_Z are étale over U, the group G acts freely on $U_{\Omega,\wp}$. We decompose $U_{Z,\wp}$ as the image of $U_{\Omega,\wp}$ and its complement. We count the number of elements of $U_{Z,\wp}$ up to $O(\mathbf{N}\wp^{n-1/2})$. The image of $U_{\Omega,\wp}$ in $U_{Z,\wp}$ is represented by the H-orbits and hence has cardinality $\sim \frac{1}{|H|}\mathbf{N}\wp^n$, and then the complement has cardinality $\sim (1 - \frac{1}{|H|})\mathbf{N}\wp^n$. Therefore the image of $U_{\Omega,\wp}$ in $U_\wp$ has cardinality $\sim \frac{1}{|G|}\mathbf{N}\wp^n$, and hence the image of $U_{Z,\wp}$ in $U_\wp$ has cardinality at most

$$\sim (\frac{1}{|G|} + 1 - \frac{1}{|H|})\mathbf{N}\wp^n.$$

In conclusion, the image of $U_{Z,\wp}$ in $U_\wp$ has cardinality

$$\leq (1 - \frac{r-1}{r|H|})\mathbf{N}\wp^n + O(\mathbf{N}\wp^{n-1/2}),$$

with $|G| = r|H|$. As $r \geq 2$ and $|G| \leq r!$, we have

$$\frac{r-1}{r|H|} \geq \frac{1}{r|H|} \geq \frac{1}{|G|} \geq \frac{1}{r!}.$$

This proves theorem 5 and hence theorem 1.

Example. Take the equation $\theta^3 = T$, with $K = \mathbf{Q}$, $K_\pi = \mathbf{Q}(\sqrt[3]{1})$, $n = 1$, $r = 3$;

if $p \equiv 1 \bmod 3$, that is to say p splits completely in K_π, the image mod p is formed of the cubes in $\mathbf{F}_p$, therefore has $(p-1)/3 + 1$ elements;

if $p \equiv -1 \bmod 3$, all the elements of $\mathbf{F}_p$ are cubes, and the image mod p has p elements (such p's are of no use for the sieve).

Exercise. Apply the sieve method of §13.2 to the example above.

(One gets the - disappointing - result that in $\mathbf{Z}$ the number of positive cubes $\leq N$ is $O(N^{1/2})$.)

13.4. Proof of theorem 3 from theorem 1.

We shall use an auxiliary result on the representations of points of $\mathbf{P}_n(K)$.

Let K be a number field, and n an integer ≥ 1.

Proposition. *There is a constant $c = c(K,n)$ such that every point $x \in \mathbf{P}_n(K)$ is representable by coordinates $(x_0, ..., x_n)$ with $x_j \in O_K$ and*

$$\left\lceil x_j \right\rceil \leq cH(x), \text{ for all } j.$$

The statement is trivial over $\mathbf{Q}$ with $c = 1$: we select the $x_i \in \mathbf{Z}$ coprime; then

$$H(x) = \sup |x_i|.$$

In the general case, note first that if $(x_0, ..., x_n)$ are coordinates of $x \in \mathbf{P}_n(K)$ with $x_i \in O_K$, the divisor class of the ideal generated by these coordinates $\underline{a}_x = (x_0, ..., x_n)$ depends only on x (and not on the chosen coordinates). We then take integral ideals $\underline{a}_1, ..., \underline{a}_h$ representing the different classes; by applying a homothety, we may assume that $x_0, ..., x_n$ generate one of the ideals $\underline{a}_i$, $(1 \leq i \leq h)$. We may then write (see §2.2(b))

$$H(x) = c_i \prod_{v \text{ arch}} \sup_{j=0,...,n} ||x_j||_v$$

$$= c_i \prod_{v \text{ arch}} \sup_{j=0,...,n} |x_j|_v^{N_v/d},$$

with $d = [K : \mathbf{Q}]$, $c_i = 1/\mathrm{N}(\underline{a}_i)^{1/d}$, and

$$N_v = \begin{cases} 1 & \text{if } v \text{ is real} \\ 2 & \text{if } v \text{ is complex} \end{cases}$$

Therefore

$$H(x) = c_i \prod_{v \text{ arch}} h_v(x),$$

where $h_v(x) = \sup_j |x_j|_v^{N_v/d} > 0$.

Lemma. *There is a constant $c'_K > 1$, depending only on K, such that for every family h_v (v runs over the set Σ of archimedean places of K) of numbers > 0, there is a a unit ϵ of K and a real number $t > 0$ with*

$$c'^{-1}_K \frac{h_v}{t} \leq ||\epsilon||_v \leq c'_K \frac{h_v}{t}.$$

We first prove this lemma. We want

$$\log ||\epsilon||_v = \log h_v - \log t + O(1),$$

that is to say

$$\log t + \log ||\epsilon||_v = \log h_v + O(1).$$

This follows from the compactness of the quotient of $\mathbf{R}^\Sigma$ by the subgroup generated by the logarithms of the units and the diagonal. One can make explicit c'_K as a function of the degree and the regulator of K (cf. [Gy]).

We now use this lemma with $h_v = h_v(x)$: if ϵ is the unit obtained in this way, we multiply x_i by ϵ^{-1}; if h is one of the $h_v(x)$ we have

$$H(x) \sim h^{(\sum N_v)/d} \sim h,$$

as $\sum N_v = d$.

This proves the proposition, since $\lceil x_j \rceil \leq \sup_v h_v(x) \sim h$.

To prove theorem 3, we consider a thin set $M \subset \mathbf{P}_n(K)$, and we define

$$M' = \{x \in O_K^{n+1}; x \neq 0, \text{the image of } x \text{ in } \mathbf{P}_n(K) \text{ belongs to } M\}.$$

Then M' is thin in $\mathbf{A}_{n+1}$. The proposition above shows that

$$M_K(X) \leq M'(cX),$$

and theorem 1 gives

$$M'(cX) = O(X^{(n+1-1/2)d}(\log X)^\gamma),$$

whence theorem 3.

APPENDIX: THE CLASS NUMBER 1 PROBLEM AND INTEGRAL POINTS ON MODULAR CURVES.

[This appendix gives a brief account of the relation between imaginary quadratic fields of class number 1 and integral points on certain modular curves. Many details have been left out, especially on the modular side. To do otherwise would have taken us too far afield and, besides, more complete results are now known, thanks to the work of Goldfeld [Gol] and Gross-Zagier [GZ].]

A.1. Historical remarks.

There are nine imaginary quadratic fields $\mathbf{Q}(\sqrt{d})$, $d < 0$, having class number equal to 1:

$$-d = 3, 4, 7, 8, 11, 19, 43, 67, 163.$$

Denote by d the discriminant of such a quadratic field K_d, $d < 0$, and $h(d)$ the class number of K_d, that is to say the number of elements of the group $\mathrm{Cl}(R_d) = \mathrm{Pic}(R_d)$, where R_d is the ring of integers of K_d.

Tables show that $h(d)$ tends to infinity as $d \to -\infty$. Gauss posed the problem of proving this (and of proving that only the nine above have $h = 1$).

In 1913, Gronwall showed that if L_d (the L-function attached to the primitive character of $(\mathbf{Z}/d\mathbf{Z})^*$ of order 2) has no zero between 1 and $1 - a/\log|d|$, (a is a given positive constant), then

$$h(d) \geq c(a)|d|^{1/2}/\log|d|(\log\log|d|)^{1/2}.$$

In particular, if GRH is true for L_d, then

$$h(d) \gg |d|^{1/2}/\log|d|,$$

(the $(\log\log|d|)^{1/2}$ term was eliminated by Landau in 1918). For the reverse inequality, it is not hard to show that (see Exercise below)

$$h(d) \leq 2|d|^{1/2}\log|d|.$$

In 1934-35, Mordell and Deuring showed that if the Riemann Hypothesis (for the usual zeta function) is false, then $h(d) \to \infty$ with $|d|$ [The idea is that if $h(d)$ is small then L_d has a zero near 1; but the existence of a zero for one L_d function entails the non-existence of zeros for the others; hence if the L_d have such zeros then ζ has none.]

Later, Heilbronn showed the same result without assuming the Riemann Hypothesis.

In 1935, Siegel proved

$$h(d) \gg |d|^{\frac{1}{2}-\epsilon}, \text{ for all } \epsilon > 0,$$

therefore

$$\log h(d) \sim \frac{1}{2}\log|d|.$$

His proof is ineffective (it uses two values of d not satisfying the desired inequality).

Heilbronn and Linfoot refined this and showed that there is at most one value of d with $|d| > 163$ and $h(d) = 1$. Landau showed more: given h, there is an explicit constant

$c(h)$ such that $h(d) \leq h$ entails $|d| \leq c(h)$ except for at most one value of d. One can choose $c(h) = 2100h^2(\log 13h)^2$. In his thesis in 1966, Stark proved that if there is a tenth imaginary quadratic field having class number 1, its discriminant d_{10} satisfies

$$|d_{10}| \geq \exp(2.2 \times 10^7) \geq 10^{9,000,000}$$

Further progress came from modular functions. In 1952, Heegner [He] published a paper proving $h(d) = 1 \Rightarrow -d \in \{3, 4, 7, 8, 11, 19, 43, 67, 163\}$ and hence solving the problem of the "tenth imaginary quadratic field". But his proof was not detailed enough to be entirely convincing (and indeed was not accepted at the time). In 1967, Stark gave a complete proof, apparently different (there are no explicit modular functions) but in fact substantially equivalent.

Also in 1967, Baker, by means of his explicit lower bounds for linear forms in logarithms, showed that $h(d) = 1 \Rightarrow |d| < 10^{500}$. (Indeed, Gel'fond and Linnik had proved in 1948 that, if $h(d) = 1$, then a certain expression of the form $|\beta_1 \log \alpha_1 + \beta_2 \log \alpha_2 + \beta_3 \log \alpha_3|$ is very small cf. [Ba] and §A.7 below.)

The problem of $h(d) = 2$ was resolved between 1972 and 1975 by Baker, Stark, Montgomery-Weinberger. By a variant of the method of Gel'fond and Linnik, they obtained an effective but very large bound for $|d|$: $|d| < 10^{1100}$, and to descend to $|d| \leq 427$ required more hard work.

In 1976, D. Goldfeld [Do1] showed that, if a certain L-function has a zero of order ≥ 4 at $s = 1$ (as it should, by the Birch and Swinnerton-Dyer conjecture), then $h(d) \to \infty$ with $|d|$, in an effective way.

[The desired property of this L-function has now been proved by Gross-Zagier [GZ], thus settling the whole problem. For instance,

$$\begin{aligned} h(d) = 3, d < 0 &\Rightarrow \log|d| \leq 21000 \\ h(d) = 4, d < 0 &\Rightarrow \log|d| \leq 33600 \\ h(d) = 5, d < 0 &\Rightarrow \log|d| \leq 35000, \end{aligned}$$

see [Go2], [Oe2].]

Exercise. Let χ_d be the quadratic character mod $|d|$ associated with the imaginary quadratic field K_d. One has

$$\chi_d(x) = -\chi_d(y) \text{ if } x \equiv -y \pmod{|d|}.$$

If $|d| > 4$, it is known ([BSh], V) that

$$h(d) = 2|d|^{1/2} L(1, \chi_d), \text{ where } L(1, \chi_d) = \sum_{n=1}^{\infty} \frac{\chi_d(n)}{n}.$$

(i) If j is an integer with $1 \leq j < |d|/2$, put:

$$l_j = \sum_{n \equiv \pm j \ (\mathrm{mod}|d|), n \geq 1} \frac{\chi_d(n)}{n} = \chi_d(j)\left(\frac{1}{j} - \frac{1}{|d| - j} + \frac{1}{|d| + j} - \frac{1}{2|d| - j} + \ldots\right).$$

Show that $|l_j| \leq 1/j$ and that $L(1, \chi_d) = \sum_{1 \leq j < |d|/2} l_j$.

(ii) Show that $\sum_{1\le j<D/2} 1/j \le \log D$ for every integer $D \ge 1$.
(iii) Deduce from (i) and (ii) that $h(d) \le 2|d|^{1/2}\log|d|$.

A.2. Equivalent conditions for $h(-p) = 1$.

The ring of integers R_d of $K_d = \mathbf{Q}(\sqrt{d})$ has $\{1, \omega_d\}$ for a basis (if d is the discriminant) where

$$\omega_d = \begin{cases} \sqrt{d}/2 & \text{if } d \equiv 0 \pmod 4 \\ (1+\sqrt{d})/2 & \text{if } d \equiv 1 \pmod 4 \end{cases}$$

The elements of order 2 of $\mathrm{Pic}(R_d)$ are easily determined. They exist if and only if d is divisible by two distinct prime factors. Thus $h(d)$ is odd if and only if $d = -4$, or $d = -8$, or $d = -p$ with p prime and $p \equiv 3 \pmod 4$.

The fields $\mathbf{Q}(i)$, $\mathbf{Q}(\sqrt{-2})$ and $\mathbf{Q}(\sqrt{-3})$ have class numbers equal to 1. We assume from now on that p is a prime number > 3 with $p \equiv 3 \pmod 4$.

Proposition. *The following properties are equivalent:*
(i) $h(-p) = 1$,
(ii) $(\frac{l}{p}) = -1$ *for all primes* l *such that* $l < p/4$.
(ii)′ $(\frac{l}{p}) = -1$ *for all primes* l *such that* $l < \sqrt{p/3}$.
(iii) *(Assuming $p > 7$) $p \equiv 3 \pmod 8$ and $R_p - N_p = 3$, where R_p (resp. N_p) is the number of quadratic residues (resp. non-residues) in the interval* $[1, (p-1)/2]$.
(iv) *The integer* $x^2 + x + \frac{p+1}{4}$ *is prime for* $x = 0, 1, \ldots, (p-7)/4$.
(iv)′ *The same assertion with* $0 \le x \le \frac{1}{2}(\sqrt{p/3} - 1)$.

Examples. Let us use (iv)′ to find those p, $3 < p < 75$, such that $h(-p) = 1$. Let $m = (p+1)/4$. For $p < 75$, we have $\frac{1}{2}(\sqrt{p/3}) - 1) < 2$, and condition (iv)′ becomes that m and $m+2$ are prime. As $m < 18$, this gives $m = 3, 5, 11, 17$ whence $p = 11, 19, 43, 67$.

To prove that $h(-163) = 1$, it suffices to check that x^2+x+41 is prime for $x = 0, 1, 2, 3$ (which gives 41,43,47, and 53). From (iv), this property remains true for $x = 4, \ldots, 39$ (as noticed by Euler, Mém. Acad. Sci. Berlin, 1772).

We can also use (iii) to check that $h(-11) = 1$: in the interval $[1, 5]$, the quadratic residues are 1,3,4,5 and the only quadratic non-residue is 2; thus $R_{11} = 4$, $N_{11} = 1$ and $R_{11} - N_{11} = 3$.

Proof of the proposition.

(i) $\Rightarrow$ (ii). For $d = -p$, we have $(\frac{l}{p}) = -1$ if and only if l is inert in the extension $K_d/\mathbf{Q}$. Consequently, if $(\frac{l}{p}) = 1$, then the ideal (l) decomposes in K_d : $(l) = \wp.\wp'$, with $l = \mathbf{N}\wp = \mathbf{N}\wp'$. If further $h(-p) = 1$, there is $\alpha \in R_d$ with $\wp = (\alpha)$, and $\mathbf{N}\alpha = l$. We write $\alpha = x + y\sqrt{-p}$ with $x, y \in \frac{1}{2}\mathbf{Z}$. Thus $\mathbf{N}(\alpha) = x^2 + py^2 = l$. As l is not a square we have $y \ne 0$ and $l \ge py^2 \ge p/4$.

(ii) $\Rightarrow$ (ii)′ is clear.

(ii)′ $\Rightarrow$ (i). Every ideal $\underline{a}$ of R_d of norm $< \sqrt{p/3}$ is generated by an integer (of $\mathbf{Z}$), thus is principal [it suffices to see this for $\underline{a} = \wp$ prime; by (ii)′, we have $\mathbf{N}(\wp) = l^2$, thus $\wp = (l)$]. But every ideal class can be shown to contain an integral ideal of norm $\le \sqrt{|d|/3}$. Therefore $h(d) = 1$.

(iii) $\Leftrightarrow$ (i). If $p > 7$, $h(-p) = 1$ implies $p \equiv 3 \pmod 8$, see below. Conversely, if $p \equiv 3 \pmod 8$, and $p > 3$, Jacobi showed that $R_p - N_p = 3h(-p)$. (The proof is analytical;

one computes the value of the L series $\sum(\frac{n}{p})n^{-s}$ at $s = 1$, see e.g. [BSh], V.§4). Hence $h(-p) = 1 \Leftrightarrow R_p - N_p = 3$.

(ii) $\Rightarrow$ (iv). We write $m = (p+1)/4$, and $f_p(X) = X^2 + X + m$. For $0 \le x \le m-2$ we have $2 \le f_p(X) < m^2$ (if $p > 3$). If, for such an integer x, $f_p(x)$ were not prime, it would be divisible by a prime number $l < m$. Thus, $x^2 + x + m \equiv 0 \pmod{l}$, and the discriminant $1 - 4m = -p$ would be a square modulo l; therefore $(\frac{-p}{l}) = 1$, whence $(\frac{l}{p}) = 1$, which contradicts (ii).

(iv) $\Rightarrow$ (iv)$'$ is obvious.

(iv)$'$ $\Rightarrow$ (ii)$'$. If (ii)$'$ were false, there would be a prime number $l < \sqrt{p/3}$ with $(\frac{l}{p}) = 1$. We have $l \ne 2$ (otherwise $p \equiv 7 \pmod 8$ and the values of $f_p(X)$ are even, which contradicts (iv)$'$). Therefore $(\frac{-p}{l}) = 1$, and the congruence $x^2 + x + m \equiv 0 \pmod{l}$ has two solutions whose sum is $l - 1$ mod l. We can then choose a solution x with $0 \le x \le (l-1)/2$. Then $f_p(x)$ is divisible by l and $f_p(x) \ge m = \frac{p+1}{4} > l$; therefore, $f_p(x)$ is not prime. As $x \le \frac{1}{2}(\sqrt{p/3} - 1)$, this contradicts (iv)$'$.

The equivalence between the properties (i) to (iv)$'$ is therefore proved.

Suppose that $p \equiv 3 \pmod 4$ and $h(-p) = 1$. From condition (ii) we deduce that
for $p > 7$, $(\frac{2}{p}) = -1$, that is to say $p \equiv 3 \pmod 8$,
for $p > 11$, $(\frac{-p}{3}) = (\frac{3}{p}) = -1$, that is to say $p \equiv 1 \pmod 3$,
for $p > 19$, $(\frac{-p}{5}) = (\frac{5}{p}) = -1$, that is to say $p \equiv 2$ or $3 \pmod 5$.

A.3. Orders of R_d.

Recall that R_d is the ring of integers of $K_d = \mathbf{Q}(\sqrt{d})$, where d is the discriminant. An *order* of R_d is a subring of R_d of rank 2 over $\mathbf{Z}$. We can uniquely write it in the form

$$R_{d,f} = \mathbf{Z} + fR_d,$$

that is to say with a basis $(1, f\omega_d)$ over $\mathbf{Z}$, where f is an integer ≥ 1 called the *conductor*. If $f > 1$, $R_{d,f}$ is not a normal (i.e. integrally closed) ring (its normalisation is R_d). The number of elements of $\mathrm{Pic}(R_{d,f})$ is denoted by $h(df^2)$.

We are interested in the (d, f) with $f \ge 2$ for which $h(df^2) = 1$. For this we use the localization exact sequence

$$R_d^* \times (\mathbf{Z}/f\mathbf{Z})^* \to (R_d/fR_d)^* \to \mathrm{Pic}(R_{d,f}) \to \mathrm{Pic}(R_d) \to 0.$$

For $d \ne -3, -4$ one has $R_d^* = \{\pm 1\}$ and this makes the computation easy.

Examples.

1) $d = -7$ and $f = 2$. In $\mathbf{Q}(\sqrt{-7})$, we have $(2) = \wp\bar{\wp}$, and $(R_d/fR_d)^* = \mathbf{F}_2^* \times \mathbf{F}_2^* = (1)$, therefore $|\mathrm{Pic}(R_{d,f})| = |\mathrm{Pic}(R_d)| = 1$.

2) $d = -11$, $f = 2$. Then (2) is inert in $\mathbf{Q}(\sqrt{-11})$, $\mathbf{F}_4^*$ is cyclic of order 3, therefore $h(-11.2^2)$ is equal to 3.

Similar arguments give the following list:

$$\text{for } d < 0 \text{ and } f \ge 2, \text{ we have } h(df^2) = 1 \Leftrightarrow \begin{cases} d = -3 & \text{and} \quad f = 2 \text{ or } 3 \\ d = -4 & \text{and} \quad f = 2 \\ d = -7 & \text{and} \quad f = 2. \end{cases}$$

A.4. Elliptic curves with complex multiplication.

Let E be an elliptic curve over a field K of characteristic zero, whose ring of endomorphisms $\mathrm{End}(E)$ over the algebraic closure of K is not equal to $\mathbf{Z}$. In this case, $\mathrm{End}(E)$ has rank 2 over $\mathbf{Z}$, and is an order $R_{d,f}$ of an imaginary quadratic field K_d.

For a given order $R_{d,f}$, there are $h = h(df^2)$ curves E up to $\bar{\mathbf{Q}}$-isomorphism with $\mathrm{End}(E) = R_{d,f}$. (One can see this over $\mathbf{C}$ by writing E as $\mathbf{C}/\Omega$, where Ω is a projective $R_{d,f}$ module of rank 1.) Thus if $h = 1$, there is a single curve up to $\bar{\mathbf{Q}}$-isomorphism; hence we have $j(E) \in \mathbf{Q}$ (indeed, if σ is an automorphism of $\bar{\mathbf{Q}}$, $\sigma(E)$ is an elliptic curve having complex multiplication by $R_{d,f}$ therefore $j(\sigma(E)) = j(E)$). Thus we can define E over $\mathbf{Q}$, but the complex multiplication is defined only after extension to K_d.

Table of the 13 values of j corresponding to $h(df^2) = 1$, $d < 0$, $f \geq 1$, i.e. to elliptic curves over $\mathbf{Q}$ with complex multiplication.

Discriminant and conductor	j	$\gamma = j^{1/3}$	$j - 2^6 3^3$
$d = -3, f = 1$	0	0	$-2^6 3^3$
$f = 2$	$2^4 3^3 5^3$	—	$2^4 3^3 11^2$
$f = 3$	$-2^{15} 3.5^3$	—	$-3.2^6 11^2 23^2$
$d = -4, f = 1$	$2^6 3^3$	$2^2 3$	0
$f = 2$	$2^3 3^3 11^3$	$2.3.11$	$2^3 3^6 7^2$
$d = -7, f = 1$	$-3^3 5^3$	-3.5	-7.3^6
$f = 2$	$3^3 5^3 17^3$	$3.5.17$	$3^8 7.19^2$
$d = -8$	$2^6 5^3$	$2^2 5$	$2^7 7^2$
$d = -11$	-2^{15}	-2^5	$-11.2^6 7^2$
$d = -19$	$-2^{15} 3^3$	$-2^5 3$	$-19.3^6 2^6$
$d = -43$	$-2^{18} 3^3 5^3$	$-2^6 3.5$	$-43.2^6 3^8 7^2$
$d = -67$	$-2^{15} 3^3 5^3 11^3$	$-2^5 3.5.11$	$-67.2^6 3^6 7^2 31^2$
$d = -163$	$-2^{18} 3^3 5^3 23^3 29^3$	$-2^6 3.5.23.29$	$-163.2^6 3^6 7^2 11^2 19^2 127^2$

The most interesting case is that of $\mathbf{Q}(\sqrt{-p})$ with $p > 11$. We take $E \cong \mathbf{C}/R_d$ where R_d is the lattice with basis $\{1, \omega_d\}$, where $\omega_d = (1 + \sqrt{-p})/2$, $d = -p$. Then we have

$$q = e^{2\pi i \omega_d} = e^{(2\pi i - 2\pi\sqrt{p})/2} = -e^{-\pi\sqrt{p}},$$

and

$$j = \frac{1}{q} + 744 + 196884q + \dots$$
$$= -e^{\pi\sqrt{p}} + 744 + \theta e^{-\pi\sqrt{p}},$$

with $10^5 \leq \theta \leq 3.10^5$.

We shall see below that j is a rational integer (and even a cube, cf. §A.6). Thus $e^{\pi\sqrt{p}}$ is close to an integer. For example (Hermite)

$$e^{\pi\sqrt{163}} = 262537412640768743.99999999999925007...$$

If one could bound the p for which

$$||e^{\pi\sqrt{p}}|| \leq \frac{10^6}{e^{\pi\sqrt{p}}},$$

(where $\|.\|$ is the distance from the nearest integer) one would obtain directly the finiteness of the p such that $h(-p) = 1$. The preceding inequality gives

$$|\log n_p - \pi\sqrt{p}| \leq \frac{10^6}{e^{2\pi\sqrt{p}}}, \quad \text{where } n_p = 1 + [e^{\pi\sqrt{p}}],$$

but n_p is too large for Baker's inequalities to apply.

To compute j, one may use the fact that $\sqrt[3]{j}/96$ is an integer close to $-\exp(\pi\sqrt{p}/3 - \log 96)$, this last number being

$-1.00028...$	for	$p = 19$
$-10.0000028...$	for	$p = 43$
$-55.000000092...$	for	$p = 67$
$-6670.000000000006...$	for	$p = 163$

Integrality of j.

Theorem. *The j invariant of an elliptic curve with complex multiplication is an algebraic integer.*

In the case $h = 1$, as $j \in \mathbf{Q}$, we then have $j \in \mathbf{Z}$.

First proof.

We choose $r \in R_{d,f}$ whose norm is a prime number l. We then have an isogeny $r : E \to E$ whose kernel is cyclic of order l. But the invariants j, j' of two elliptic curves related by an isogeny of degree l satisfy a relation

$$F_l(j, j') = 0,$$

where F_l is a symmetric polynomial over $\mathbf{Z}$ with highest degree term $j^l j'^l$. Here $j = j'$ and

$$F_l(j, j) = j^{2l} + \sum_{i=0}^{2l-1} a_i j^i = 0, \quad (a_i \in \mathbf{Z}),$$

thus j is an algebraic integer.

Second proof.

That j is integral at a place v is equivalent to E having potential good reduction at v (that is to say, good reduction after making a finite extension of the ground field). But if E does not have potential good reduction then it has potential multiplicative reduction at v, i.e. a reduction to the group $\mathbf{G}_m$. The endomorphisms of E then operate on $\mathbf{G}_m$, which is absurd, as the ring of endomorphisms of $\mathbf{G}_m$ is $\mathbf{Z}$.

(For a generalization of this to abelian varieties, see [ST].)

Remark. If E has complex multiplication by R_d, then over $\mathbf{Q}(\sqrt{d})$, $j(E)$ generates the Hilbert class field of $\mathbf{Q}(\sqrt{d})$, cf. e.g. [Shm].

A.5. Modular curves associated to normalisers of Cartan subgroups, and their CM-integral points.

The method.

Let us explain the relation between the class number 1 problem and integral points on modular curves.

Let p be a sufficiently large prime number ($p > 19$, say) such that the imaginary quadratic field $\mathbf{Q}(\sqrt{-p})$ has class number $h(-p) = 1$. Let us consider the elliptic curve E, unique up to $\mathbf{C}$-isomorphism, which has complex multiplication by the ring of integers $R = R_{-p}$ of the field $\mathbf{Q}(\sqrt{-p})$. Its modular invariant j is in $\mathbf{Z}$, thus defines an integral point of the affine line $\mathbf{P}_1 - \{\infty\}$ which is the simplest modular curve "$X(1)^{\mathrm{aff}}$". We shall attach to E some integral points on some other modular curves. After that, there are essentially three ways to procede:

1). (Heegner, Siegel). A direct method gives all the integral points of the modular curve, thus giving an explicit list of the possible p's.

2). Baker's method is applied to bound the integral points of the modular curve, hence bounding p.

3). One obtains a linear form in logarithms for which Baker's lower bound applies. This method is closely related to the two preceding, and also to the "limit formula" of Kronecker.

Modular curves associated with Cartan subgroups and their normalisers.

Let N be an integer > 1, and $X(N)$ the modular curve which classifies elliptic curves with a basis of points of order N, see e.g. [DeRa]. The field of functions of $X(N)$ is the field of modular functions of level N rational over $\mathbf{Q}(\sqrt[N]{1})$ cf. [Shm]; its field of constants is $\mathbf{Q}(\sqrt[N]{1})$. The group $G_N = \mathrm{GL}_2(\mathbf{Z}/N\mathbf{Z})$ acts on $X(N)$, the quotient being the j-line $X(1) = \mathbf{P}_1$.

If H is a subgroup of G_N, we write X_H for the modular curve $X(N)/H$; its points over a field K correspond to elliptic curves E over K such that $\mathrm{Gal}(\bar{K}/K)$ acts on the N-division points of E via H, see [DeRa], IV-3. If det: $H \to (\mathbf{Z}/N\mathbf{Z})^*$ is surjective, X_H is absolutely irreducible over $\mathbf{Q}$.

The subgroups H we are interested in are essentially the Cartan subgroups of G_N and some subgroups of their normalisers. More precisely:

Let A be a free commutative étale $\mathbf{Z}/N\mathbf{Z}$-algebra of rank 2: if $p|N$, A/pA is either $\mathbf{F}_p \times \mathbf{F}_p$ (called the "split" case) or $\mathbf{F}_{p^2}$ (the "non-split" case at p).

Given an arbitrary decomposition of the primes $p|N$ into "split" and "non-split" types, then up to isomorphism there is a unique corresponding such algebra A. In the case where A is non-split at p for all $p|N$, we say more briefly that A is "non-split".

Let A^* be the multiplicative group of invertible elements of A; then A^* operates on A, and as $A \cong (\mathbf{Z}/N\mathbf{Z})^2$, A^* embeds in $\mathrm{Aut}_{\mathbf{Z}/N\mathbf{Z}}(A) \cong \mathrm{GL}_2(\mathbf{Z}/N\mathbf{Z})$. The image of A^* is a Cartan subgroup C of $\mathrm{GL}_2(\mathbf{Z}/N\mathbf{Z})$.

As A is an algebra of rank 2, it has a unique automorphism σ of order 2, which induces the natural involution on each A/pA; for all a, $a + \sigma a \in \mathbf{Z}/N\mathbf{Z}$. Let $\tilde{C} = C \cup \sigma C$ be the subgroup of $\mathrm{GL}_2(\mathbf{Z}/N\mathbf{Z})$ generated by C and σ.

Let $\tilde{X} = X_{\tilde{C}}$ and $X = X_C$ be the corresponding curves. As C has index 2 in $\tilde{C}$, we have a quadratic covering $X \to \tilde{X}$.

Let now E be an elliptic curve with complex multiplication by the ring R, and suppose that N is prime to the discriminant of R. Then the algebra $A = R/NR$ is étale, thus defines a C and a $\tilde{C}$. Let K be a field over which E is defined, and suppose that $\sqrt{d} \notin K$. Let E_N be the kernel of multiplication by N in $E(\bar{K})$; this is a free R/NR-module of rank 1. If $s \in \mathrm{Gal}(\bar{K}/K)$ is trivial on $K(\sqrt{d})$, then the action of s on E_N commutes with that of R/NR, hence is defined by an element of the Cartan subgroup $C = (R/NR)^*$. If s is non-trivial on $K(\sqrt{d})$, its action on E_N comes from an element of $\tilde{C} - C$. This allows us to attach to E a K-rational point of $\tilde{X}$, which does not lift to a point of X over K, but which lifts over $K(\sqrt{d})$. In other words, we obtain two points of X conjugate under $\mathrm{Gal}(K(\sqrt{d})/K)$.

We now take the case which interests us: $K = \mathbf{Q}$, $R = R_{-p}$ and $h(-p) = 1$. Recall that every prime number $l < p/4$ is inert in $\mathbf{Q}(\sqrt{-p})$. In particular if p is large with respect to N, then C and $\tilde{C}$ are of non-split type.

We then obtain rational points on suitable modular curves $\tilde{X}$. These points are *integral points* in the following sense:

First, by removing the cusps from a modular curve X_H, one obtains an affine curve X_H^{aff}, which is finite over the j-line $\mathbf{A}^1$:

$$\begin{array}{ccc} X_H^{\mathrm{aff}} & \to & X_H \\ \downarrow & & \downarrow j \\ \mathbf{A}^1 & \to & \mathbf{P}_1. \end{array}$$

By taking the normalisation of $\mathbf{A}^1_{\mathbf{Z}}$, one then obtains an affine scheme $X_{H,\mathbf{Z}}^{\mathrm{aff}}$ over $\mathbf{Z}$. A rational point x of X_H is then called *integral* if it has the following equivalent properties:

a) $j(x)$ is integral,

b) x is an integral point of the scheme $X_{H,\mathbf{Z}}^{\mathrm{aff}}$.

The points $x_E \in \tilde{X}(\mathbf{Q})$ constructed above have property a) by §A.4: they are indeed *integral.*

The case where N is prime.

Assume $N = l$ is a prime number ≥ 5 and consider the curve $\tilde{X}$ corresponding to the normaliser $\tilde{C}$ of the non-split Cartan subgroup C.

Since C has order $l^2 - 1$, it has index $l(l-1)$ in $\mathrm{GL}_2(\mathbf{Z}/l\mathbf{Z})$, thus $X \to \mathbf{P}_1$ has degree $l(l-1)$ and $\tilde{X} \to \mathbf{P}_1$ has degree $l(l-1)/2$. The number of geometric cusps of X (resp. $\tilde{X}$) is $l-1$ (resp. $(l-1)/2$). All these cusps have ramification index l. The $l-1$ cusps of X are conjugate over $\mathrm{Gal}(\mathbf{Q}(\sqrt[l]{1})/\mathbf{Q})$ and the $(l-1)/2$ cusps of $\tilde{X}$ are conjugate over the maximal real subfield of the cyclotomic field $\mathbf{Q}(\sqrt[l]{1})$.

By working out the ramification above $j = 0, 1728$ one obtains a formula for the genus $\tilde{g}$ of $\tilde{X}$ (in the non-split case):

$$\tilde{g} = \begin{cases} (l^2 - 10l + 33)/24 & \text{if} \quad l \equiv 1 \bmod 12 \\ (l-5)^2/24 & \text{if} \quad l \equiv 5 \bmod 12 \\ (l-3)(l-7)/24 & \text{if} \quad l \equiv 7 \bmod 12 \\ (l^2 - 10l + 13)/24 & \text{if} \quad l \equiv 11 \bmod 12 \end{cases}$$

Examples. $l = 5$, $\tilde{g} = 0$, 2 cusps; $l = 7$, $\tilde{g} = 0$, 3 cusps; $l = 11$, $\tilde{g} = 1$, 5 cusps.

Note that, already for $l = 7$ and $l = 11$, the number of integral points of $\tilde{X}$ is finite (and effectively so by Baker's method, once explicit equations for $\tilde{X}$ have been found). This is enough to give, by method 2, that the set of p's with $h(-p) = 1$ can be effectively bounded.

A.6. Examples.

In what follows, p is a prime number > 3 with $h(-p) = 1$; we denote by j_p the j-invariant of an elliptic curve with CM by R_{-p} i.e.

$$j_p = j(q_p), \text{ where } q_p = e^{2\pi i(1+\sqrt{-p})/2} = -e^{-\pi\sqrt{p}}.$$

(For a model of this curve with good reduction outside p, see [Gr], p.80.) We have $j_p \in \mathbf{Z}$ cf. §A.4. By using the modular curves $\tilde{X}$ associated to $N = 2, 3, \ldots$ we will obtain further properties of j_p (and hence of p).

N = 2. The group $\mathrm{GL}_2(\mathbf{Z}/2\mathbf{Z})$ is isomorphic to S_3. The non-split Cartan subgroup C is the cyclic subgroup of order 3, and $\tilde{C} \cong S_3$, therefore $\tilde{X} = \mathbf{P}_1$ and X is a 2-sheeted covering of $\mathbf{P}_1$. By looking at its ramification, and at the cusps, one can see that $X \to \tilde{X} = \mathbf{P}_1$ is the quadratic extension associated to $\sqrt{j - 1728}$.

Theorem. *If $h(-p) = 1$ and $p \equiv 3 \pmod 8$, then $j_p - 1728 = -py^2$ with $y \in \mathbf{Z}$.*

(Indeed, the corresponding point of X is rational over $\mathbf{Q}(\sqrt{-p})$.) This can also be checked using the classical formulae

$$j - 1728 = c_6^2/\Delta \text{ and } \Delta^{1/2} = 4(e_1 - e_2)(e_2 - e_3)(e_3 - e_1).$$

N = 3. Consider the natural homomorphisms

$$\mathrm{GL}_2(\mathbf{F}_3) \to \mathrm{PGL}_2(\mathbf{F}_3) = S_4 \to S_3.$$

The group $\tilde{C}$ is a 2-Sylow subgroup corresponding to the inverse image of a subgroup of order 2 of S_3. The covering $\tilde{X} \to \mathbf{P}_1$ has degree 3, and becomes cyclic over $\mathbf{Q}(\sqrt{-3})$. One shows that it is obtained by extracting the cube root of j. As above this implies:

Theorem. *If p is such that 3 is inert in $\mathbf{Q}(\sqrt{-p})$, then j is a cube: $j = \gamma^3$, $\gamma \in \mathbf{Z}$.*

The result is again true when 3 splits, for the normaliser of a split Cartan subgroup is contained in the normaliser of a non-split Cartan subgroup. (Here again an elementary proof is possible by using an explicit formula for $\Delta^{1/3}$ see e.g. [Se6], §5.3(b).)

Remark. If we write $j_p - 1728 = -py^2$ and $j_p = \gamma^3$, we obtain by putting $X = -p\gamma$, $Y = p^2 y$, and $k = 2^6 3^3 p^3$,

$$Y^2 - X^3 = k.$$

As

$$\gamma = -e^{\pi\sqrt{p}/3}(1 + O(e^{-\pi\sqrt{p}})),$$

X and Y have order of magnitude the exponential of $\sqrt{p} \sim k^{1/6}$. This alone should be enough to imply a bound for p [it would if even a weak form of the "ABC-conjecture" were true].

Returning to the case $N = 3$, if one uses the 2-sheeted covering $X \to \tilde{X}$, one finds that

$$-3((\frac{\gamma}{12})^2 + \frac{\gamma}{12} + 1) = -pz^2, \text{ with } z \in \mathbf{Z}.$$

As $j_p - 1728 = -py^2$, and $j_p = \gamma^3$, putting $u = y/24z$ we have

$$\frac{\gamma}{12} - 1 = -u^2.$$

A priori $u \in \mathbf{Q}$. In fact, one can show that $u \in \mathbf{Z}$ and u is divisible by 3:

$p=$	19	43	67	163
$u=$	±3	±9	±21	±231

N = 5. The curve $\tilde{X}$ has genus 0 and 2 cusps rational over $\mathbf{Q}(\sqrt{5})$. An explicit parametrisation of it is given in [Si4]. The corresponding integral points can be identified with the units of $\mathbf{Q}(\sqrt{5})$.

N = 6. The equation $\gamma = 12(1-u^2)$ given above corresponds in fact to the equation of $\tilde{X}$ for $N = 6$.

N = 7. One has $\tilde{g} = 0$, and there are 3 cusps, conjugate over the real cyclotomic subfield of $\mathbf{Q}(\sqrt[7]{1})$ (cubic field).

One finds that $h(-p) = 1$, and $(\frac{p}{7}) = -1$, gives an exceptional unit of that field. See [Ke2]. There are analogous results for $N = 9$.

N = 12. One again finds a curve of genus 0 with 2 cusps which are rational over $\mathbf{Q}(\sqrt{3})$. One can either use the method of approximating $e^{\pi\sqrt{p}}$, or pass to level 24.

N = 15. This is the level used by Siegel [Si4]. An elliptic curve is obtained, with a finite number of integral points.

N = 24. An elliptic curve is obtained. This is the level considered in effect by Heegner [He].

A.7. The Gel'fond-Linnik-Baker method.

Let N be the discriminant of a real quadratic field - this field is then $\mathbf{Q}(\sqrt{N})$. Let $\tilde{X}$ be the curve attached to the non-split Cartan subgroup $\tilde{C}$ relative to N. There is (see below) a non-trivial morphism ϕ over $\mathbf{Q}$ of $\tilde{X}$ onto a curve of genus 0, such that the image of $\tilde{X}^\infty$ is formed of two points P,P' conjugate to each other over the field $\mathbf{Q}(\sqrt{N})$, and the inverse image of $\{P, P'\}$ is $\tilde{X}^\infty$. (For $N = 5$ one can select ϕ to be an isomorphism.) Over $\mathbf{Q}(\sqrt{N})$ one can interpret ϕ as a function $\tilde{X} \to \mathbf{P}_1$ having neither zeros nor poles over $\tilde{X}^{\mathrm{aff}}$ and with $\phi.\phi^\sigma = $ cst., where σ is the involution of $\mathbf{Q}(\sqrt{N})$.

Here is a first method for showing the existence of such a function, in case N is a prime l with $l \equiv 1 \pmod 4$. Consider the group of divisor classes generated by the differences of two cusps. By general results of Drinfeld-Manin and Kubert-Lang, this group is finite. The cusps are conjugate among themselves under the Galois group $(\mathbf{Z}/l\mathbf{Z})^*/(\pm1)$. As the Legendre character is trivial on ±1, it gives a homomorphism $\epsilon : (\mathbf{Z}/l\mathbf{Z})^*/(\pm1) \to (\pm1)$. Since ϵ has order 2, it gives a decomposition of the cusps into two subsets of $(l-1)/4$ elements each. For a suitable integer $a \geq 1$, one can thus construct a function whose divisor is a times an element of the first set, minus a times an element of the second. One obtains thus the desired function ϕ; that it is defined over $\mathbf{Q}(\sqrt{N})$ derives from its uniqueness (up to a constant factor).

Classical formulae provide another method for constructing ϕ. For instance, in case N is a prime $l \equiv 1 \pmod 4$, one looks at the action of the Cartan subgroup C on the projective line $\mathbf{P}_1(\mathbf{F}_l)$.

The cyclic group $C/\mathbf{F}_l^*$ acts freely and transitively on $\mathbf{P}_1(\mathbf{F}_l)$; its unique subgroup of index 2 has two orbits A and B. To each $x \in \mathbf{P}_1(\mathbf{F}_l)$, let us associate the modular form Δ_x defined by

$$\Delta_\infty = \Delta(lz),$$

$$\Delta_i(z) = \Delta(\frac{z+i}{l}), \quad i \in \mathbf{Z}/l\mathbf{Z},$$

where Δ is the standard discriminant. The function ϕ given by

$$\phi = \frac{\prod_{a\in A} \Delta_a}{\prod_{b\in B} \Delta_b},$$

can be shown to have the required properties. (One can sometimes use $\eta = \Delta^{1/24}$ instead of Δ e.g. for $l = 5$, cf. [Si4]; this gives a smaller order at infinity.)

From the existence of ϕ, one deduces a bound for the p for which $h(-p) = 1$ in the following way. We may assume that p is sufficiently large so that we are in the non-split case (every prime divisor l of N is $< p/4$). To p is associated an integral point x_p of $\tilde{X}$. The image $\phi(x_p)$ of x_p belongs to a quasi-integral subset of $\mathbf{P}_1 - \{0, \infty\}$. More concretely, we can write

$$\phi(x_p) = c\epsilon_p,$$

where ϵ_p is a unit, and c belongs to a fixed set, independent of p. Let $\epsilon_N > 1$ be a fundamental unit of $\mathbf{Q}(\sqrt{N})$. It suffices to consider a sequence of p for which $\phi(x_p) \to +\infty$, thus $\epsilon_p = \pm\epsilon_N^{n_p}$ where n_p is a sequence of integers tending to $+\infty$. Let us look at the size of $\phi(x_p)$.

As ϕ and j are related by a polynomial equation we have a Puiseux expansion

$$\phi = \lambda j^\alpha(1 + O(j^{-\beta})),$$

where α and β are rational numbers > 0, $\lambda \neq 0$. As

$$j = e^{\pi\sqrt{p}}(-1 + O(e^{-\pi\sqrt{p}})),$$

we obtain

$$c\epsilon_N^{n_p} = \pm\phi(x_p) = \pm\lambda e^{\alpha\pi\sqrt{p}}(1 + O(e^{-\nu\pi\sqrt{p}})),$$

where ν is a rational number > 0. Then

$$\left|\frac{\pm c\epsilon_N^{n_p}}{\lambda e^{\alpha\pi\sqrt{p}}} - 1\right| \ll e^{-\nu\pi\sqrt{p}}.$$

In particular, n_p has order of magnitude $\sqrt{p}$. We thus obtain a linear form in 3 logarithms, to which one can apply Baker's lower bounds.

This had been seen by Gel'fond and Linnik in a different manner. They looked at the extensions

$$\begin{array}{ccccc} & & \mathbf{Q}(\sqrt{N},\sqrt{-p}) & & \\ & \nearrow & \uparrow & \nwarrow & \\ \mathbf{Q}(\sqrt{N}) & & \mathbf{Q}(\sqrt{-Np}) & & \mathbf{Q}(\sqrt{-p}) \\ & \nwarrow & \uparrow & \nearrow & \\ & & \mathbf{Q} & & \end{array}$$

and considered the L-function attached to the character χ of the imaginary quadratic field $\mathbf{Q}(\sqrt{-p})$ corresponding to its quadratic extension $\mathbf{Q}(\sqrt{N},\sqrt{-p})$. We have

$$(1) \qquad L(s,\chi) = L(s,\chi_N)L(s,\chi_{-Np}),$$

where χ_N (resp. χ_{-Np}) is the Dirichlet character corresponding to $\mathbf{Q}(\sqrt{N})/\mathbf{Q}$ (resp. $\mathbf{Q}(\sqrt{-Np})/\mathbf{Q}$)). In particular:

$$L(1,\chi) = L(1,\chi_N).L(1,\chi_{-Np}).$$

The numbers $L(1,\chi_N)$ and $L(1,\chi_{-Np})$ are $\log \epsilon_N$ and $h(-Np)$, up to elementary factors. As for $L(1,\chi)$, one computes it using Kronecker's limit formula applied to the expansion

$$L(1,\chi) = \frac{1}{2}\sum_{x,y\in\mathbf{Z}} \chi_N(F_p(x,y))F_p(x,y)^{-1},$$

where

$$F_p(x,y) = x^2 + xy + \frac{p+1}{4}y^2.$$

One finds that $L(1,\chi)$ is essentially $\log \phi(x_p)$. Using (1), this gives the expression of $\phi(x_p)$ as $\pm c\epsilon_N^{n_p}$ as above (and gives the values of c and n_p). For more details see e.g. [Si3].

Baker chose $N = 21$; then $\epsilon = \epsilon_N = \frac{1}{2}(5+\sqrt{21})$ and one finds

$$|h(-21p)\log\epsilon_N - \frac{32}{21}\pi\sqrt{p}| < e^{-\pi\sqrt{p}/100}.$$

Note that these linear forms in logarithms have algebraic numbers as coefficients (for example $\frac{32}{21}i\sqrt{p}$), so that Baker's method does apply, and gives an effective bound for p.

Remarks. 1) Note the connection between logarithms of units and values of L-functions at $s = 1$! It has been the starting point of Stark's conjectures, cf. [St] and [T3].
2) It is not known how to solve the class number 2 problem by methods of integral points on modular curves. But this problem can be attacked via linear forms in logarithms in the following way.

Suppose $h(-d) = 2$. The difficult case is that where $d = p_1p_2$, with p_1 and p_2 large prime numbers, $p_1 \equiv 1 \pmod 4$, $p_2 \equiv 3 \pmod 4$. One knows that the corresponding j-invariant belongs to $\mathbf{Q}(\sqrt{p_1})$. Then $\phi(x_d)$ is essentially a unit in $\mathbf{Q}(\sqrt{N},\sqrt{p_1})$. Only 2 of the 3 fundamental units of this biquadratic field are involved, namely ϵ_N and ϵ_{Np_1}. By working this out, Baker obtains [Ba, p.53]:

$$|h(-21d)\log\epsilon_N + h(21p_1)h(-21p_2)\log\epsilon_{21p_1} - \frac{64}{21}\pi\sqrt{d}| < e^{-\sqrt{d}/10}.$$

As ϵ_{21p_1} is a fairly large algebraic number, one must use delicate lower bounds to be able to conclude.

Bibliography.

[A] S. ARAKELOV - Intersection theory of divisors on an arithmetic surface (in Russian), Izv. Akad. Nauk SSSR 8 (1974), 1167-1180.

[Ba] A. BAKER - *Transcendental number theory*, Cambridge Univ. Press, Cambridge, 1975.

[Be] G.V. BELYI - On the Galois extensions of the maximal cyclotomic field (In Russian), Izv. Akad. Nauk. SSSR 43 (1979), 267-276.

[Ber] D. BERTRAND - Lemmes de zéros et nombres transcendants, Séminaire Bourbaki 1985-86, exposé 652 (Astérisque 145-146, S. M. F. 1987, 21-44).

[BM] A. BAKER and D.W. MASSER (edit.) - *Transcendence Theory: Advances and Applications*, Acad. Press, London, 1977.

[Bo] F.A. BOGOMOLOV - Sur l'algébricité des représentations l-adiques, C.R. Acad. Sci. Paris, série A, 290 (1980), 701-703.

[Bom1] E. BOMBIERI - *Le grand crible dans la théorie analytique des nombres,* Astérisque 18 (1974), 2nd ed. (1987).

[Bom2] E. BOMBIERI - On exponential sums over finite fields II, Invent. Math. 47 (1978), 20-39.

[Bor] A. BOREL - Some finiteness properties of adèle groups over number fields, I.H.E.S. Publ. Math. 16 (1963), 5-30 (= *Oe.*, II, no. 60, p.329-354).

[BSD] B.J. BIRCH and H.P.F. SWINNERTON-DYER - Notes on elliptic curves, I, J. Crelle 212 (1963), 7-25; II, *ibid.*, 218 (1965), 79-108.

[BSh] Z. BOREVICH and I. SHAFAREVICH - *Number Theory*, Acad. Press, New York, 1966.

[Ca] J.W.S. CASSELS - Diophantine equations with special reference to elliptic curves, J. London Math. Soc. 41 (1966), 193-291.

[Ch] C. CHABAUTY - Sur les points rationnels des courbes algébriques, C.R.Acad. Sci. Paris 212 (1941), 882-885.

[CN] K. CHANDRASEKHARAN and R. NARASIMHAN - Functional equations with multiple gamma factors and the average order of arithmetical functions, Ann. of Math. 76 (1962), 93-136.

[Co] S.D. COHEN - The distribution of Galois groups and Hilbert's irreducibility theorem, Proc. London Math. Soc. 43 (1981), 227-250.

[Col] R. COLEMAN - Effective Chabauty, Duke Math. J. 52 (1985), 765-770.

[Cu] G.V. CHUDNOVSKY - Formules d'Hermite pour les approximants de Padé de logarithmes et des fonctions binômes, et mesures d'irrationalité, C. R. Acad. Sci. Paris, Ser. A, 288 (1979), 965-967.

[CW] C. CHEVALLEY and A. WEIL - Un théorème d'arithmétique sur les courbes algébriques, C.R. Acad. Sci. Paris 195 (1932), 570-572. (= A. Weil, *Oe.*, I [1932c]).

[D] V.A. DEMJANENKO - Rational points of a class of algebraic curves (in Russian), Izv. Akad. Nauk SSSR 30 (1966), 1373-1396.

[De] H. DELANGE - Généralisation du théorème de Ikehara, Ann. Sci. ENS 71 (1954), 213-242.

[DeRa] P. DELIGNE and M. RAPOPORT - Les schémas de modules de courbes elliptiques, Lect. Notes in Math. 349, Springer-Verlag, 1973, 143-316.

[Dr] V.G. DRINFELD - Two theorems on modular curves (in Russian), Funk. Anal. i Prilozhen. 7 (1973), 83-84.

[DR] H. DAVENPORT and K.F. ROTH - Rational approximation to algebraic numbers, Mathematika 2 (1955), 160-167. (= H. Davenport, *Coll. Works*, II, p.689-696).

[E] T. EKEDAHL - An effective version of Hilbert's irreducibility theorem, preprint, Stockholm, 1987.

[F] G. FALTINGS - Endlichkeitssätze für abelsche Varietäten über Zahlkörpern, Invent. Math. 73 (1983), 349-366.

[Fe] D. FENNA - Rational approximation with series, J. Austral. Math. Soc. 2 (1961/2), 107-119.

[FJ] M.D. FRIED and M. JARDEN - *Field Arithmetic*, Springer-Verlag, 1986.

[Fr] M. FRIED - Constructions arising from Néron's high rank curves, Trans. AMS 281 (1984), 615-631.

[FW] G. FALTINGS, G. WÜSTHOLZ et al. - *Rational Points*, Vieweg, Braunschweig, 1984.

[G] A. GROTHENDIECK - *Éléments de Géométrie Algébrique* (written with the collaboration of J. Dieudonné), I.H.E.S. Publ. Math. 1960-67.

[GH] P. GRIFFITHS and J. HARRIS - *Principles of Algebraic Geometry*, Wiley, 1978.

[Go1] D. GOLDFELD - The class numbers of quadratic fields and the conjectures of Birch and Swinnerton-Dyer, Ann. Sci. Norm. Sup. Pisa 3 (1976), 623-663.

[Go2] D. GOLDFELD - Gauss' class number problem for imaginary quadratic fields, Bull. AMS 13 (1985), 23-37.

[Gr] M. GREENBERG - Rational points in Henselian discrete valuation rings, I.H.E.S. Publ. Math. 31 (1966), 56-64.

[Gy] K. GYÖRY - Résultats effectifs sur la représentation des entiers par des formes décomposables, Queen's Papers in Pure and Applied Math., no.56 (1980).

[GZ] B.H. GROSS and D.B. ZAGIER - Heegner points and derivatives of L-series, Invent. Math. 84 (1986), 225-320.

[H] R. HARTSHORNE - *Algebraic Geometry*, GTM 52, Springer-Verlag, 1977.

[Ha] M. HARRIS - Systematic growth of Mordell-Weil groups of abelian varieties in towers of number fields, Invent. Math. 51 (1979), 123-141.

[He] K. HEEGNER - Diophantische Analysis und Modulfunktionen, Math. Zeit. 56 (1952), 227-253.

[HH] D. HILBERT and A. HURWITZ - Über die diophantischen Gleichungen vom Geschlecht Null, Acta Math. 14 (1891), 217-224. (= D. Hilbert, *Ges. Abh.*, Bd. 2, 258-263 = A. Hurwitz, *Math. Werke*, Bd.2, 116-121).

[Hi] D. HILBERT - Über die Irreducibilität ganzer rationaler Funktionen mit ganzzahligen Koeffizienten, J. Crelle 110 (1892), 104-129 (= *Ges. Abh.*, II, 264-286).

[Hn1] M. HINDRY - Points de torsion sur les sous-variétés abéliennes, C. R. Acad. Sci. Paris 304 (1987), Série 1, 311-314.

[Hn2] M. HINDRY - Géométrie et hauteurs dans les groupes algébriques Université de Paris VI, Thèse, 1987.

[Ho] T. HONDA - Isogenies, rational points and section points of group varieties, Jap. J. Math. 30 (1960), 84-101.

[Hu] M.N. HUXLEY - The large sieve inequality for algebraic number fields, Mathematika 15 (1968), 178-187.

[Hup] B. HUPPERT - *Endliche Gruppen I*, Springer-Verlag, 1967.

[HW] G.H. HARDY and E.M. WRIGHT - *An introduction to the theory of numbers*, 5th ed. Oxford Univ. Press, London 1979.

[Is] V.V. ISHANOV - The semi-direct imbedding problem with a nilpotent kernel, Izv. Akad. Nauk SSSR 40 (1976), 3-25 (= Math. USSR Izv. 10 (1976), 1-23).

[K] S. KLEIMAN - Toward a numerical theory of ampleness, Amer. J. Math. 84 (1966), 293-344.

[Ke1] M.A. KENKU - On the number of **Q**-isomorphism classes of elliptic curves in each **Q**-isogeny class, J. Number Theory 15 (1982), 199-202.

[Ke2] M.A. KENKU - A note on the integral points of a modular curve of level 7, Mathematika 32 (1985), 45-48.

[KL] D. KUBERT and S. LANG - *Modular Units*, Springer-Verlag, Heidelberg 1981.

[L1] S. LANG - *Abelian Varieties*, Interscience Publ., New York, 1959.

[L2] S. LANG - *Diophantine Geometry*, Interscience Publ., New York, 1962.

[L3] S. LANG - Les formes bilinéaires de Néron et Tate, Sém. Bourbaki 1963/64, exposé 274, Benjamin Publ., New York, 1966.

[L4] S. LANG - Division points on curves, Ann. Mat. Pura Appl. IV, 70 (1965), 229-234.

[L5] S. LANG - *Elliptic Curves-Diophantine Analysis*, Springer-Verlag, Heidelberg, 1978.

[L6] S. LANG - *Fundamentals of Diophantine Geometry*, Springer-Verlag, Heidelberg, 1983.

[L7] S. LANG - Integral points on curves, I.H.E.S. Publ. Math. 6 (1960), 27-43.

[La] E. LANDAU - Über die Anzahl der Gitterpunkte in gewisser Bereichen, Gött. Nach. (1912), 687-771, and (1915), 209-243.

[LaM] S. LAMACCHIA - Polynomials with Galois group PSL(2, 7), Comm. in Algebra 8 (1980), 983-992.

[LN] S. LANG and A. NÉRON - Rational points of abelian varieties over function fields, Amer. J. Math. 81 (1959), 95-118.

[Lau] M. LAURENT - Equations diophantiennes exponentielles, Invent. Math. 78 (1984), 299-327.

[Len] H.W. LENSTRA, Jr. - Rational functions invariant under a finite abelian group, Invent. Math. 25 (1974), 299-325.

[Lg] M. LANGEVIN - Quelques applications de nouveaux résultats de van der Poorten, Séminaire Delange-Pisot-Poitou 17 (1976), no. G12.

[Lev] J. LEVEQUE - Rational points on curves of genus 1, J. Crelle 206 (1961), 45-52.

[LU] R.C. LYNDON and J.L. ULLMAN - Groups generated by two parabolic linear fractional transformations, Canad. J. Math. 21 (1969), 1388-1403.

[LW] S. LANG and A. WEIL - Number of points on varieties in finite fields, Amer. J. Math. 76 (1954), 819-827 (= A. Weil, *Oe.*, II, [1954f]).

[M] L.J. MORDELL - On the rational solutions of the indeterminate equations of the third and fourth degrees, Proc. Cambridge Phil. Soc. 21 (1922), 179-192.

[Ma1] Y. MANIN - The p-torsion of elliptic curves is uniformly bounded (in Russian), Izv. Akad. Nauk SSSR 33 (1969), 459-465 (= Math. USSR Izv. 3 (1969), 433-438).

[Ma2] Y. MANIN - Cyclotomic fields and modular curves (in Russian), Usp. Mat. Nauk 26 (1971), 7-71 (= Russian Math. Surveys 26 (1971), 7-78).

[Ma3] Y. MANIN - Parabolic points and zeta functions of modular curves (in Russian), Izv. Akad. Nauk SSSR 36 (1972), 19-66 (= Math. USSR Izv. 6 (1972), 19-64).

[Mac] A. MACINTYRE - Non-standard number theory, Proc. International Congress Math. Helsinki, Vol. 1 (1978), 253-262.

[Mas1] D.W. MASSER - Diophantine approximation and lattices with complex multiplication, Invent. Math. 45 (1978), 61-82.

[Mas2] D.W. MASSER - Small values of the quadratic part of the Néron-Tate height, Séminaire Delange Pisot Poitou, 1979-1980, Birkhäuser-Verlag, 1981, 213-222.

[Mat] B.H. MATZAT - *Konstruktive Galoistheorie*, Lect. Notes in Math. 1284, Springer-Verlag, 1987.

[Me] J-F. MESTRE - Formules explicites et minorations de conducteurs de variétés algébriques, Comp. Math. 58 (1986), 209-232.

[MFT] Y. MANIN, J. FRANKE, Y. TSCHINKEL - Rational points of bounded height on Fano varieties, Invent. Math. 1988.

[Mi] M. MIGNOTTE - Approximation des nombres algébriques, Publ. Math. Orsay, no. 77.74 (1977).

[MM] G. MALLE and B.H. MATZAT - Realisierung von Gruppen $PSL_2(\mathbf{F}_p)$ als Galoisgruppen über $\mathbf{Q}$, Math. Ann. 272 (1985), 549-565.

[Mo] H.L. MONTGOMERY - The analytic principle of the large sieve, Bull. AMS 84 (1978), 547-567.

[Mu1] D. MUMFORD - A remark on Mordell's conjecture, Amer. J. Math. 87 (1965), 1007-1016.

[Mu2] D. MUMFORD - *Abelian Varieties*, 2nd ed., Tata Inst., Oxford Univ. Press, Oxford, 1974.

[Mz1] B. MAZUR - Rational points of abelian varieties with values in towers of number fields, Invent. Math. 18 (1972), 183-266.

[Mz2] B. MAZUR - Rational isogenies of prime degree, Invent. Math. 44 (1978), 129-162. (See also Sém. Bourbaki 1977/78, exposé 511.)

[N1] A.NÉRON - Problèmes arithmétiques et géométriques rattachés à la notion de rang d'une courbe algébrique dans un corps, Bull. Soc. Math. France 80 (1952), 101-166.

[N2] A. NÉRON - Propriétés arithmétiques de certaines familles de courbes algébriques, Proc. Int. Cong. Amsterdam, 1954, vol.III, 481-488.

[N3] A. NÉRON - Arithmétique et classes de diviseurs sur les variétés algébriques, Proc. Int. Symp. Tokyo-Nikko, 1955, 139-154.

[N4] A. NÉRON - Quasi-fonctions et hauteurs sur les variétés abéliennes, Ann. of Math. 32 (1965), 249-331.

[Na] K. NAKATA - On some elliptic curves defined over **Q** of free rank ≥ 9, Manus. Math. 29 (1979), 183-194.

[No] D.G. NORTHCOTT - An inequality in the theory of arithmetic on algebraic varieties, Proc. Cambridge Phil. Soc. 45 (1949), 502-509; A further inequality ..., ibid., 510-518.

[Ne] J. NEUKIRCH - On solvable number fields, Invent. Math. 53 (1979), 135-164.

[Oe1] J. OESTERLÉ - Nombres de Tamagawa et groupes unipotents en caractéristique p, Invent. Math. 78 (1984), 13-88.

[Oe2] J. OESTERLÉ - Le problème de Gauss sur le nombre de classes, L'Ens. Math., to appear.

[Oe3] J. OESTERLÉ - Nombres de classes des corps quadratiques imaginaires, Sém. Bourbaki no.631, Astérisque 121-122, SMF, 1985.

[P] H. POINCARÉ - Sur les propriétés arithmétiques des courbes algébriques, J. de Liouville 7 (1901), 166-233 (= *Oe.*, V, 483-548).

[Pa] A.N. PARSHIN - Algebraic curves over function fields I (in Russian), Izv. Akad. Nauk SSSR 32 (1968), 1191-1219.

[Ra] D.A. RAIKOV - A generalization of the Ikehara-Landau theorem (in Russian), Math. Sbornik 45 (1938), 559-568.

[R1] M. RAYNAUD - Courbes sur une variété abélienne et points de torsion, Invent. Math. 71 (1983), 207-233.

[R2] M. RAYNAUD - Around the Mordell conjecture for function fields and a conjecture of Serge Lang, Lect. Notes in Math. 1016, Springer-Verlag, (1983), 1-19.

[R3] M. RAYNAUD - Sous-variétés d'une variété abélienne et points de torsion, *Arithmetic and Geometry*, Progress in Math. 35, Birkhäuser-Verlag, Boston (1983), 327-352. (See also Sém. Bourbaki 1983-84, exposé 625.)

[Rob] A. ROBINSON and P. ROQUETTE - On the finiteness theorem of Siegel and Mahler concerning Diophantine equations, J. Number Theory 7 (1975), 121-176.

[Ro] P. ROQUETTE - Non standard aspects of Hilbert's irreducibility theorem, Lect. Notes in Math. 498, Springer-Verlag, 1975, 231-275.

[Sch] S.H. SCHANUEL - Heights in number fields, Bull. Soc. Math. France 107 (1979), 433-449.

[Schi] A. SCHINZEL - On Hilbert's irreducibility theorem, Acta Arith. 16 (1965), 334-340.

[Schm1] W.M. SCHMIDT - Approximation to algebraic numbers, L'Ens. Math. 17 (1971) 187-253. (Monographie no. 19, 1972).

[Schm2] W.M. SCHMIDT - *Diophantine approximation,* Lecture Notes in Math. 785, Springer-Verlag, 1980.

[Schu] I. SCHUR - Gleichungen ohne Affekt, Sitz. Preuss. Akad. Wiss. 1930, Physik.-Math. Klasse 443-449 (= *Ges. Abh.* III, 191-197).

[Se1] J-P. SERRE - Quelques applications du théorème de densité de Chebotarev, Publ. Math. I.H.E.S. 54 (1981), 123-201.

[Se2] J-P. SERRE - Quelques propriétés des groupes algébriques commutatifs, Astérisque, 69-70 (1979), 191-202.

[Se3] J-P. SERRE - *Abelian l-adic representations and elliptic curves*, Benjamin, New York 1968.

[Se4] J-P. SERRE - Divisibilité de certaines fonctions arithmétiques, L'Ens. Math. 22 (1976), 227-260.

[Se5] J-P. SERRE - Groupes de Galois sur **Q**, Séminaire Bourbaki 1987, exposé 689.

[Se6] J-P. SERRE - Propriétés galoisiennes des points d'ordre fini des courbes elliptiques, Invent. Math. 15 (1972), 259-331.

[Sh1] K-y. SHIH - On the construction of Galois extensions of function fields and number fields, Math. Ann. 207 (1974), 99-120.

[Sh2] K-y. SHIH - p-division points on certain elliptic curves, Comp. Math. 36 (1978), 113-129.

[Sha] I.R. SHAFAREVICH - Construction of fields of algebraic numbers with given solvable Galois group, Izv. Akad. Nauk SSSR 18 (1954), 525-578. (AMS Transl. Ser. 2, 4 (1956), 107-142.)

[Shm] G. SHIMURA - *Introduction to the arithmetic theory of automorphic functions*, Publ. Math. Soc. Japan, vol. 11, Princeton Univ. Press, 1971.

[Si1] C.L. SIEGEL - Über einige Anwendungen diophantischer Approximationen, Abh. Pr. Akad. Wiss. (1929), no.1 (=*Ges. Abh.*, I, 209-266).

[Si2] C.L. SIEGEL - Abschätzung von Einheiten, Gött. Nach. (1969), 71-86 (= *Ges. Abh.*, IV, 66-81.)

[Si3] C.L. SIEGEL - *Lectures on advanced analytic number theory*, Tata Institute, Bombay 1965.

[Si4] C.L. SIEGEL - Zum Beweise des Starkschen Satzes, Invent. Math. 5 (1968), 180-191 (=*Ges. Abh.*, IV, 41-52).

[Sk] T. SKOLEM - *Diophantische Gleichungen*, Springer-Verlag, 1938 (Chelsea Publ. Comp., New York, 1950).

[Sp] V.G. SPRINDZUK - Hilbert's irreducibility theorem and rational points on algebraic curves, Soviet Math. Dokl. 20 (1979), 701-705.

[St] H. STARK - Values of L-functions at $s = 1$, I,II,III,IV; Adv. in Math. 7 (1971), 301-343; 17 (1975), 60-92; 22 (1976), 64-84; 35 (1980), 197-235.

[ST] J-P. SERRE and J. TATE - Good reduction of abelian varieties, Annals of Math. 78 (1968), 492-517.

[Sw] H.P.F. SWINNERTON-DYER - *Analytic theory of abelian varieties*, London Math. Soc. Lecture Notes 14, C.U.P. (1974).

[Sz] L. SZPIRO - *Séminaire sur les pinceaux arithmétiques: la conjecture de Mordell*, Astérisque 127 (1985), S.M.F.

[T1] J. TATE - The arithmetic of elliptic curves, Invent. Math. 23 (1974), 179-206.

[T2] J. TATE - Quasifonctions locales, letters to J-P. Serre, 21 June 1968, and 1st October 1979 (see [L5], [Z]).

[T3] J. TATE - *Les conjectures de Stark sur les fonctions L d'Artin en* $s = 0$, Progress in Math. 47, Birkhäuser-Verlag, Boston, 1984.

[Ta] T. TATUZAWA - On a theorem of Siegel, Jap. J. of Math. 21 (1951), 163-178.

[Ti] R. TIJDEMAN - Exponential diophantine equations, Proc. Inter. Congress Math. Helsinki, Vol. 1 (1978), 381-387.

[To] J. TOP - Néron's proof of the existence of elliptic curves over **Q** with rank at least 11, preprint, Utrecht 1987.

[U] S. UCHIYAMA - On the Thue-Siegel-Roth theorem, III, Proc. Japan Acad. 36 (1960), 1-2.

[Wa1] M. WALDSCHMIDT - *Nombres transcendants et groupes algébriques*, Astérisque 69-70 (1979).

[Wa2] M. WALDSCHMIDT - A lower bound for linear forms in logarithms, Acta. Arith. 37 (1980), 257-283.

[W1] A. WEIL - L'arithmétique sur les courbes algébriques, Acta Math. 52 (1928), 281-315 (= *Oe.*, I, [1928]).

[W2] A. WEIL - Sur un théorème de Mordell, Bull. Sc. Math. 54 (1929), 182-191 (=*Oe.*, I, [1929]).

[W3] A. WEIL - *Variétés abéliennes et courbes algébriques*, Hermann, Paris, 1948.

[W4] A. WEIL - Remarques sur un mémoire d'Hermite, Arch. d. Math. 5 (1954), 197-202 (= *Oe.*, II, [1954a]).

[Wh] E.T. WHITTAKER and G.N. WATSON - *A course of modern analysis*, 4th edition, C.U.P. (1969).

[Z] H.G. ZIMMER - Quasifunctions on elliptic curves over local fields, J. Crelle 307/308 (1979), 221-246; Corrections, *ibid.* 343 (1983), 203-211.

[Za] D. ZAGIER - L-series of elliptic curves, the Birch - Swinnerton- Dyer conjecture, and the class number problem of Gauss, Notices AMS 31 (1984), 739-743.

INDEX

Aspects of Mathematics

English-language subseries (E)

Aspekte der Mathematik

Deutschsprachige Unterreihe (D)

GPSR Compliance

The European Union's (EU) General Product Safety Regulation (GPSR) is a set of rules that requires consumer products to be safe and our obligations to ensure this.

If you have any concerns about our products, you can contact us on ProductSafety@springernature.com

In case Publisher is established outside the EU, the EU authorized representative is:

Springer Nature Customer Service Center GmbH
Europaplatz 3
69115 Heidelberg, Germany

Batch number: 10577876

Printed by Printforce, the Netherlands